T0177593

THE DEATH OF THE ETHIC OF LIFE

THE DEATH OF THE ETHIC OF LIFE

John Basl

OXFORD
UNIVERSITY PRESS

Oxford University Press is a department of the University of Oxford. It furthers
the University's objective of excellence in research, scholarship, and education
by publishing worldwide. Oxford is a registered trade mark of Oxford University
Press in the UK and certain other countries.

Published in the United States of America by Oxford University Press
198 Madison Avenue, New York, NY 10016, United States of America.

© Oxford University Press 2019

All rights reserved. No part of this publication may be reproduced, stored in
a retrieval system, or transmitted, in any form or by any means, without the
prior permission in writing of Oxford University Press, or as expressly permitted
by law, by license, or under terms agreed with the appropriate reproduction
rights organization. Inquiries concerning reproduction outside the scope of the
above should be sent to the Rights Department, Oxford University Press, at the
address above.

You must not circulate this work in any other form
and you must impose this same condition on any acquirer.

Library of Congress Cataloging-in-Publication Data
Names: Basl, John, author.
Title: The death of the ethic of life / John Basl.
Description: New York, NY : Oxford University Press, [2019] |
Includes bibliographical references and index.
Identifiers: LCCN 2018026316 (print) | LCCN 2018039342 (ebook) |
ISBN 9780190923907 (online content) | ISBN 9780190923884 (updf) |
ISBN 9780190923891 (epub) | ISBN 9780190923877 (cloth : alk. paper)
Subjects: LCSH: Life. | Ethics.
Classification: LCC BD435 (ebook) | LCC BD435.B385 2019 (print) |
DDC 179/.1–dc23
LC record available at https://lccn.loc.gov/2018026316

9 8 7 6 5 4 3 2 1

Printed by Sheridan Books, Inc., United States of America

CONTENTS

ACKNOWLEDGMENTS

This book is as it is because of the care, attention, love, guidance, and influence of many. While I am to fully blame for its faults, its virtues, such as they are, should be credited to these others as much as to myself.

I have been thinking, at least somewhat carefully, about the issues at the heart of this work for about a decade. In that time, I have been fortunate to share my ideas and have them shaped and improved by many philosophers. I'm especially cognizant of the role that two philosophers have played in the development of the arguments of this book and in my development as a philosopher: Ron Sandler and Elliott Sober. I have been arguing with Ron about issues of moral status, environmental ethics, and many other topics since I was, almost literally, a child. He has talked me out of more bad views and bad arguments than I can count, certainly more than I care to admit in public. He has been a constant source of encouragement, advice, and support, and his mark on my philosophical thinking is indelible. Elliott is one of the most insightful and clearheaded philosophical thinkers I have ever met. My arguments with Elliott have been among the most rewarding and humbling of my philosophical career so far, and they will not be easily surpassed. It was through discussions with Elliott that I was inspired to take up this topic and to draw so heavily on the tools and work in philosophy of biology to address issues of moral status. Ron and Elliott are both philosophers worth emulating, but more importantly, they are people worth emulating. I am forever grateful for their mentorship.

Various arguments or their ancestors that appear throughout this work have been improved by feedback from the very attentive audiences at various workshops, conferences, colloquia, etc. In 2012, I had the great pleasure of working with Sune Holm to organize a workshop on teleological organization held at the Carlsberg Akademi in Copenhagen. With generous funding from the Danish Council for Independent Research, we were able to bring together a wonderful group, and I benefited greatly from the workshop. I'd like to thank not only Sune, but also Karen Neander, Jay Odenbaugh, Fred Bouchard, Samir Okasha, Elselijn Kingma, Arno Wouters, Simon Rippon, and Wybo Houkes. After the workshop, Sune and I collected contributions from the workshop, as well as others, and pitched a special issue on teleological organization to *Synthese*. I'm grateful to Sune and those at *Synthese*, especially Otávio Bueno, for making that project possible, as well as to two anonymous reviewers for helpful feedback in developing my own contribution to that issue.

In 2013, I presented a first version of the arguments that would become the "Trilemma for Teleological Individualism" at the Rocky Mountain Ethics Congress. RoME has always been an excellent venue for me to get feedback on my work and to develop lasting friendships and working relationships with philosophers, and this was no exception. I'm especially grateful to Ben Hale for his comments on my paper and to Mylan Engel Jr. for his feedback.

In 2015, Greg Mikkelson organized the Workshop on Value Theory in Environmental Ethics, at the Centre de Recherche in Ethique at the University of Montreal, to which I was invited to give a talk on some of the key arguments in this book. The group, which included Greg, Gary Varner, John Nolt, Antoine Dussault, and Valéry Giroux was fantastically helpful. Their feedback and interest has been both helpful and motivating as I worked to complete this project.

Ron Sandler and I have coauthored two papers on the moral status of synthetic organisms, and I draw on many of the arguments developed in those papers in this book. One paper was published in *Studies in History and Philosophy of Science Part C: Studies in History and Philosophy of Biological and Biomedical Sciences*, the other in *Synthetic Biology and Morality*. In both instances, we received very helpful feedback that much improved the arguments. My thanks to the various anonymous referees that provided such feedback and to Gregory Kaebnick and Thomas Murray, the editors for the collected volume, who also provided fantastic feedback.

I was also fortunate to have received extremely helpful reviews from two reviewers at Oxford University Press, and even more fortunate to have the support and aid of Lucy Randall and her assistant Hannah Doyle at OUP in shaping and guiding this project in myriad ways. I'm also indebted to Richard Isomaki for his careful copy-editing of the manuscript.

There are of course many others who have asked questions, provided feedback, pushed me on points, and helped me to develop my views and arguments. I'm sorry to anyone who has offered feedback at these sorts of venues whom I have failed to mention.

I owe a huge debt of gratitude to the departments and institutions that have supported me and my work. As an undergraduate, I was adopted by the

Department of Philosophy and Religion at Northeastern University. It was my home away from home, and the faculty there saw me through my philosophical infancy. In 2013, some of these very same faculty became my colleagues when I joined the department as a member of the faculty. There were, and continue to be, many new faces. While some of my former mentors have since retired, the department continues to help me flourish both philosophically and personally, and I'm very grateful to both my former mentors and my current colleagues. I want to especially thank Steve Nathanson, Liz Bucar, Serena Parekh, and Candice Delmas for their relentless encouragement and offers of assistance as I wrote this book. This work has been improved greatly by my being able to trouble Rory Smead, and his regular collaborator Patrick Forber, with questions about technical aspects of evolutionary biology. I'm thankful for their expertise and their friendship. Ben Yelle has also helped me greatly in thinking through the issues of welfare developed here. In addition to my department, the College of Social Sciences and Humanities and Northeastern University as a whole have been very generous in support of my work.

Before coming to Northeastern, I was fortunate to have spent two years as a faculty member at Bowling Green State University. Many of the arguments in this book were honed by discussions with my faculty colleagues there, and equally so by my graduate student colleagues. I'm especially indebted to Christian Coons. Working with him has been one of the highlights of my career, and his friendship one of the highlights of my adult life. In addition to the BGSU community as a whole, I'd like to particularly thank Michael Bradie, Margy DeLuca, Priscilla Ibarra, Lou Katzner, Marcus Schultz-Bergin, Arthur Ward, Michael Weber, Mark Wells, and Sara Worley, and the students in my Biocentrism seminar in the spring of 2013 for working through these issues in detail.

This project started as a dissertation at University of Wisconsin–Madison and would not have been possible without the help of so many members of the UW community. It was a wonderful place to work on a project that draws on resources from ethics and philosophy of biology. In addition to Elliott, I'd like to especially thank Jim Anderson, Rob Streiffer, Dan Hausman, and Mike Titelbaum for their contributions to this project.

At every institution at which I've been a part, I've had the opportunity to teach classes that engage with the core issues of this project. I've been lucky to have classes full of engaged students who have helped me to stay excited about these issues and remind me why they matter and why I care about them. To all my students, thank you. I owe a special debt to four students in particular for their help in preparing this manuscript: Sam Haas, Myraeka d'Leeuwen, Gabriel Morris, and Davi Sherman.

Finally, there are the friends and family who have served as my support network, sounding board, Socratic partner, and stress relief. I am extremely fortunate to have some people in my life that are not just fantastic philosophers but dear friends: people with whom it is just as easy to slip into philosophical discussion as it is to slip out of it. Among these are Matt Kopec, Jeff Behrends, Gina Schouten, Matt Barker, Ben Miller, Mike Roche, Mike Goldsby, and Danielle Wylie. To these and many others, thank you all for doing philosophy with me and for not doing philosophy with me. Just as essential to completing this project were the friends who aren't philosophers, though most are philosophical. I'm especially grateful for the friendships of Paul Giangregorio, Rob DiPietro, Chadi Salamoun, Bo Kim-Kopec, and Erin Barker. Often while writing this book and needing a break, I would decide to take my dog, Arya, for a walk and watch her play with her friends in the neighborhood. Arya has good tastes in dogs and people alike, and I was lucky to have the regular companionship of Carlos, Dianne, Jeff, Doreen, and all their pups. And Arya herself has played a special role in keeping me grounded while working through this project and through much else.

I count myself lucky to have a family that has always been supportive, that never tried to convince me that it was a bad idea to go into philosophy, encouraging me in my chosen career. Thank you, Mom, Dad, and family. To my brother, Josh, and his partner, Keryn, and to my grandparents, Mary, John, and Roz, thank you for always making room in your lives for me, for making your home my home, and for much, much more.

Last, but very far from least, I would not be the philosopher, or the person, that I am today without the love and support of Evelyn Eastmond. She has been with me through so much of my life that it is hardly possible, even as a philosopher, to imagine what life might have been like without her. I am certain that without her, this book, nor any other by me, would exist. I am deeply indebted to her for the motivation, care, love, support, and inspiration she provided. Thank you, Evelyn.

THE DEATH OF THE ETHIC OF LIFE

INTRODUCTION

According to the ethic of life, living things are of special moral importance; they are due respect, concern, care, or reverence for their own sake in virtue of the fact that they are alive. Living things are not like road signs or rocks. An ethic of life makes no distinction, at least no categorical one, between those living things that are conscious or sentient and those that are not. Humans, elephants, fish, and ferns differ in many important ways, but, according to the ethic of life, each is a moral subject, one that we, as agents, ought to be responsive to independently of whether we take a liking to, or otherwise depend on, these subjects.

The ethic of life takes many forms and has a place in many ethical and religious traditions. Traces of such an ethic can be found in Jainism and in some Native American traditions. It would be surprising if the ethical commitments of many conservationists and environmentalists weren't shaped, at least implicitly, by a commitment to the moral value of living things. The ethic of life is perhaps most well known to philosophers as the view of Albert Schweitzer, from whom the name of the ethic is drawn. Schweitzer saw living things not as mere complex systems, but as each possessing a will striving toward life. He developed his ethic of reverence for life on the basis of this commitment. In his words:

> Ethics consists, therefore, in my experiencing the compulsion to show to all will-to-live the same reverence as I do my own. There we have given us that basic principle of the moral which is a necessity of thought. It is good to maintain and encourage life; it is bad to destroy life or to obstruct it. (Schweitzer 1969, 309)

Who or what matters from the moral point of view is a central theoretical question in environmental ethics in the Western academic tradition. There is, I would wager, no class taught on environmental ethics that does not engage this question, attempting to discern, in some fashion, whether our environmental use or treatment is limited only by a concern for humans or whether our moral concerns do or ought extend further to sentient beings, to all organisms, or even to species and ecosystems themselves.

In the Western academic tradition, the ethic of life is known as *biocentrism* or *biocentric individualism*. Biocentrism is not just a vague commitment to the moral value of life; it is an ethic of life that has been clarified, refined, revised. Biocentrism is constituted by a nexus of claims that philosophers have worked hard to substantiate, dedicating serious intellectual effort to crafting arguments for the view and developing frameworks for deriving our obligations in light of the commitment to the moral value of all living things. They have taken pains to distinguish living things from other things that lack such importance, things such as ecosystems and species, rocks and road signs. It is this ethic of life that the title of this book refers to. This book is an argument that biocentrism, an ethic on which all living beings have special moral value, is mistaken and must be abandoned.

Among environmental ethicists who think about moral status or moral standing, a comfortable stalemate holds between the biocentrists and sentientists, who believe that some form of consciousness or sentience marks the boundaries of moral concern. Biocentrists and sentientists agree, in broad strokes, about so much. They both see an intimate connection between welfare and moral status, and they both adopt similar strategies in defending their preferred moral boundaries. They both agree that holism is false and for mostly the same reasons. The points of disagreement between the parties typically concern how we are to understand welfare or well-being, what capacities of a thing are essential to having interests. Here each party has a repertoire of stock arguments. The sentientist says that having a welfare requires having interests, and having interests requires cognitive capacities; the biocentrist articulates a distinction between *taking an interest*, which does require cognitive capacities, and *having an interest*, which does not. The sentientist argues that to the extent that nonsentient organisms have interests, those interests are derivative; the biocentrist brings forth examples where the relevant interests cannot be derivative—and so on and so forth.

My aim is to break this stalemate, to argue for an old conclusion in a new way. In doing so, I aim to not only undermine biocentrism, but to expose some deep flaws in the views of sentientists concerning the nature of welfare, to provide a clear framework for thinking about issues of moral status, and to provide the resources for parties to these debates to understand the way that issues in the philosophy of biology and philosophy of technology bear on these debates. The constellation of views I defend will, I am sure, not sit well with most parties to the debates over moral status. They include the following:

- Nonsentient organisms have interests, a welfare, or good of their own.
- Some nonsentient biological collectives have interests, a welfare, or a good of their own.
- Species and ecosystems are not among the collectives that have interests, a welfare, or good of their own.
- Artifacts, even simple artifacts, have interests, a welfare, or good of their own . . . in the same sense as nonsentient organisms.
- Not all welfare matters from the moral point of view, including the welfare of nonsentient organisms, collectives, and artifacts.

Endorsing the conjunction of these views is heterodox, and making the case for it requires engaging carefully with questions about the nature of welfare, moral status, naturalized teleology, the levels of selection, and the workings of emerging technology. My arguments draw on work on all of these issues, and I hope to show that careful attention to them does more than show that we must accept the death of the ethic of life.

Chapter Summaries

The book is divided into two parts. Part I consists of three chapters.

Chapter 1 articulates the commitments of biocentrism vis-à-vis explaining the form of moral status that advocates of the view take living things to have, moral considerability, as well as the strategies these advocates employ both for arguing that all living things are morally considerable, and for excluding certain things, such as artifacts and ecosystems, from being morally considerable. I identify an approach to arguments over issues of moral status that is common to both sentientists and biocentrists. I call this the *welfare approach*. Those working in the welfare approach see moral status as being grounded in welfare or well-being; arguments over who or what has moral status, as well as arguments for what things are to be excluded from having moral status, are taken as debates over which things bear or have a welfare. I argue that biocentrists are bound to at least some, if not all, of the components of the welfare approach; they cannot adopt alternative views about moral status, such as those advocated by some holists, or defend their views using those strategies. This will be unobjectionable to most biocentrists since they have often criticized those alternative approaches. However, I will ultimately show that there is an unresolvable tension in biocentrists' attempts to mark the boundaries of moral considerability. The resources that they must appeal to in order to make sense of the moral considerability of all living things make it impossible to truly rule out nonsentient collectives and artifacts.

Chapter 2 takes up two distinct sets of challenges to biocentrism. The first concerns the relationship between moral status and normative theory. The challenge is that questions of moral status are sensitive to and settled by questions of which normative theory is true, and so there is no defending the claim that nonsentient organisms have moral status without defending a particular normative theory. I articulate several different questions about moral status, questions about the bearers of moral status, questions about the implications of having moral status, and questions about the grounds for moral status. I explain the ways in which these questions are and are not sensitive to normative theory, and defend the view that questions about the bearers of moral status can be settled independently of issues of normative theory. The second challenge, the *subjectivist challenge*, rests on the claim that there is no satisfactory account of welfare that does not depend in some way on the bearer of welfare having cognitive capacities, that attributions of welfare to nonsentient things are illusory, derivative, etc. Here I make space for the welfare of nonsentient organisms by defending an objective-list view of welfare and using the subjectivist challenge to set conditions of adequacy for a theory of welfare for nonsentient organisms.

Chapter 3, the final chapter of Part I, develops an *etiological account of teleological welfare*, an account that satisfies the conditions of adequacy set forth in the previous chapter and so answers the subjectivist challenge. I explain that *if* nonsentient organisms really are teleologically organized, their good can be defined in terms of their ends in a way that is nonarbitrary, nonderivative, and subject relative. However, this depends on providing a naturalized account of teleology—one on which teleology isn't merely illusory, arbitrary, or derivative. Borrowing insights from etiological theories of function, I develop an etiological account of teleology, explaining why it is superior to a theory that simply defines welfare in terms of functions. I also argue that the etiological account of teleology is really the only game in town as far as biocentrism is concerned; alternative accounts of naturalized teleology, such as autopoietic accounts, are ill-suited to the aims of defending biocentrism or grounding teleological welfare.

Part I makes space for biocentrism; it develops the resources for biocentrists to make good on their claim that nonsentient organisms have welfare in a real and genuine sense. Part II makes the case that both nonsentient artifacts and collectives have a welfare in that same sense, and that there is no plausible way to draw boundaries around the morally considerable in a way that biocentrism requires.

Chapter 4 begins to develop the case against biocentrism. The problem for biocentrism is that, by definition, it is *only* organisms that are morally considerable. The view is inherently antiholist and antiartifact. Once we adopt the etiological account of teleological welfare as the best and only plausible account of the welfare of nonsentient organisms, biocentrism faces a problem of exclusion; it is not possible to adopt the etiological account of teleology in a way that grounds the welfare of nonsentient organisms while excluding biological collectives or artifacts. Chapter 4 develops this problem of exclusion with respect to biological collectives. This requires careful consideration of an issue within the philosophy of biology: the problem of the levels or units of selection. I argue that the biocentrist is committed to adopting a view about the levels of selection that grounds teleological individualism, the view that only individual organisms are teleologically organized. But among the views available concerning which things are ultimately subject to natural selection, none of them justify teleological individualism in a way that satisfies other biocentric commitments. I then argue that the correct view about the units of selection is one on which biological collectives are sometimes teleologically organized, and I explore which collectives might be so organized, and so which things those that accept an etiological theory of teleological welfare must recognize as having a welfare.

Chapter 5 develops the problem of exclusion with respect to artifacts. It seems clear that artifacts are teleologically organized. Biocentrists, and anyone else interested in allowing that nonsentient organisms have a welfare while denying that artifacts have one, must find some way to distinguish the teleological organization of artifacts from that of organisms. I argue that artifacts are teleologically organized in the same way as nonsentient organisms, and that both have a welfare in the same way. The argument for this proceeds by considering the variety of ways that advocates of teleological welfare have attempted to defend an asymmetry between artifacts and organism, including appeals to the derivative nature of artifact teleology, the mind-dependence of artifact teleology, the living/nonliving distinction, and the natural/artificial selection distinction. I argue that none of these attempts succeed; they either fail to draw the relevant line between even simple artifacts and nonsentient organisms, or they are arbitrary, ad hoc, or question-begging. Many have seen such an implication as a *reductio* of biocentrism, but I don't believe that this is so. I explain why, even though biocentrism is false, it shouldn't be dismissed simply because we must countenance artifact welfare.

Chapter 6 considers what options are left for the biocentrist. In my view, if one is committed to biocentrism, one ought really be committed to a view that I will call teleocentrism, the view that all things that are teleologically organized have a welfare. I explain why this view is false, and so biocentrism is also false. I first argue that there are no grounds for accepting that the welfare of nonsentient organisms is morally significant while denying the significance of the same kind of welfare in nonorganisms, such as artifacts and biological collectives. In other words, if biocentrists wish to maintain that nonsentient organisms are morally considerable, they must give up biocentrism in favor of teleocentrism. I then appeal to a series of cases and arguments to show that teleocentrism is to be rejected primarily because of the way it would force us to bring artifacts into the moral fold. I argue that, on balance, the most justified position is one on which we simply accept that not all welfare matters, that teleological welfare is of no direct moral significance.

The conclusion takes up the broader implications of the death of the ethic of life in four domains: environmental ethics and environmental practice, medicine and medical ethics, emerging technologies, and within philosophy more broadly. Given the webs of interdependence in nature, I argue that not much hangs, in terms of policy, on the fact that biocentrism and teleocentrism are false, but there are edge cases: cases where, for example, we might be thought to have an obligation to restore specific species or make reparations for past environmental wrongdoing, where the answer to questions about moral considerability matters. In the domain of medicine and medical ethics, biological health is often seen as normatively significant. This is evident in, for example, cases where we deny a request to amputate a limb for someone that suffers from bodily identity integrity disorder (BIID). Such individuals do not see their healthy limb as part of them and would prefer to remove what is seen as an alien body from their own body. The fact that we deny such requests, indeed the fact that we classify BIID as a disorder at all, seems to presuppose that biological health is normative, or so I argue. I explore this case and several other implications for rejecting the moral significance of biological welfare. Third, I consider what the implications are for emerging technologies. I argue that the conclusions of this book undermine any concern we should have about how to protect nonsentient organisms that result from synthetic biology for their own sake and, similarly, undermine recent attempts to defend the rights of mere machines or autonomous computer systems that are not yet conscious. Finally, I consider how the arguments impact debates within philosophy. In particular, I explain how the arguments shift the lines of debate and open up new grounds for philosophical interventions within the domains of environmental ethics, moral status, and welfare.

Methodology

As is customary when writing a book in applied ethics, I'll say a bit here about my methodological commitments, though they are discussed and defended in more depth in the conclusion. As one might expect of a project that engages with issues in the empirical sciences and the philosophy of science and biology, and with fundamental issues in value theory, I will employ a variety of methodological tools throughout this book. The one tool that will be most polarizing is that of (wide) reflective equilibrium, a methodology that, by its nature, appeals to philosophical intuitions. I will ask the reader to consider carefully constructed thought experiments, some not merely hypothetical, but also very strange. I will ask readers to make a carefully considered judgment about such cases and will employ my own such judgments in the service of defending the conclusions of this book. That isn't to say that there will be no arguments or principles defended. The judgments about cases will be weighed against various principles and be used in support of the premises of various arguments. In some instances, I will argue in favor of overriding particular judgments in favor of principles and in the face of compelling arguments.

Most readers of this book will, I suspect, be familiar with this methodology, and I won't fully defend it.[1] Instead, I will simply lay out my hand and give a brief overview of the reasons I endorse the method that I do. I think that the method of reflective equilibrium is the best tool we have for settling various foundational matters concerning normativity (and perhaps other philosophical issues). It seems to me that there are roughly four main positions one might take about philosophical methodology as applied to identifying normative truths. The first is a type of rationalism. The sort of rationalism I'm thinking of is of the Cartesian or Kantian variety, one on which normative truths will show themselves to be true, and necessarily so, by pure reasoning, their denial will be a contradiction or, at the very least, inconceivable. On this view, the discovery of normative truths is akin to the process of proving new theorems in mathematics. There are some who still have hope that such a

---

1. The method of reflective equilibrium, as described by Rawls (1999), refers to any method on which we arrive at our final views about both moral judgments in particular cases and moral principles by weighing and balancing them against one another. Wide reflective equilibrium (see Daniels 1979) adds in further judgments or facts from other domains that serve as constraints on the weighing and balancing of moral principles against moral judgments. I don't mean for my use of reflective equilibrium to perfectly represent the model as described by either Rawls or Daniels, and I don't intend it solely to apply to moral issues. Rather than describe the method in any abstract way, I'll try to explain throughout the considerations appealed to in my weighing and balancing of principles and particular judgments about cases.

project will be successful.[2] I hope they are right. The completion of this rationalist project would be an awesome intellectual achievement. But I'm not hopeful. I think, in the end, we cannot have a fully rationalist account of the domain.

The second broad methodological position is that of naturalism. The naturalist thinks that descriptive truths, discoverable by broadly empirical methods, provide the foundations for normativity.[3] For the naturalist, facts (or knowledge of facts) about the evolution of, for example, social cooperation or altruism tells us something important about what the social norms *are*.[4] For my part, I think that naturalistic methods do tell us something important about ethics and normativity. For one thing, evolution sets limits on moral requirements. We can't be morally required to do that which we are incapable of doing, and evolution influences our capacities and capabilities. But, as a general tool for determining and defending a view about the normative truths, it seems to me that naturalism isn't up to the task. The method seems straightforwardly *descriptive* rather than *prescriptive*. It tells us how the world is, what it is like, and how it came to be that way, but not whether that is desirable, good, or whether we ought to change it. Of course, this objection begs various questions against naturalists, and they have their favored responses, but ultimately, I think that there is something important about normativity that can't be explicated on a naturalist picture, and naturalists will ultimately have to deny that kind of normativity in favor of some thinner notion that they will no doubt argue is adequate to the tasks to which it is put.

A third option is a kind of pessimism or skepticism. According to such a view, if the kind of normative truths we are in search of aren't discoverable a priori and are largely immune to discovery by the methods of naturalism, so

---

2. Once I was talking to Ralf Bader, a philosopher whom I greatly admire but who is not a fan of the method of reflective equilibrium. I asked him, "What if our most well justified moral principles tell us that we are morally required to kick others in the face?" to which he replied, with a wry smile, "Get good boots."

3. This form of naturalism should not be confused with metaethical naturalism, taken as the view that moral facts supervene on or are constituted by natural facts, nor with physicalism. Instead, naturalism here picks out a particular program by which we discover which natural facts determine the moral facts, and on which moral norms just are identified with the norms that have evolved or are contracted into given our social arrangements. This form of naturalism is perhaps best understood as Hobbesian.

4. There are various ways that the evolution of social cooperation might be relevant to morality, and the tools and methods used for drawing conclusions about the evolution of social cooperation are invaluable for a range of projects (see Skyrms 2014; Forber and Smead 2014). The view I wish to deny is one on which the evolutionary facts are somehow constitutive of substantive normative truths.

much the worse for such truths. We should deny that there are any such truths or, at least, think they are inaccessible to us. There are many forms of pessimism within philosophy, especially within ethics: metaethical noncognitivists, error theorists, and relativists, along with their normative counterparts. Not all pessimists are *methodological pessimists*, pessimists about arriving at normative truths via reflective equilibrium. Many pessimists are pessimists precisely because they have certain commitments that they come to via reflective equilibrium that entail their form of pessimism. Nor are all pessimists merely *moral* pessimists; some pessimists take their pessimism much further. There is no room here to respond to all these forms of pessimism and, for the most part, no need to do so.

But there are two considerations that I think speak against a pessimism about the conclusions and worth of this project. First, I think there is no room to be both a *methodological* and *purely moral* pessimist. That is, one can't think that there is no plausible method for discovering moral or ethical truths or norms alone. At the heart of all intellectual endeavors are a set of norms. At the heart of science is a commitment to some set of norms about the relationship of evidence to hypothesis, for example. These inform how we develop and interpret the results of experiments. Debates between Bayesians (Howson and Urbach 2006; Sober 2008, chap. 1) and frequentists (Mayo 1996) are normative debates. It seems to me that the same tools available for settling these debates are available for settling moral debates. If one is pessimistic about these tools in the moral case, one should be equally pessimistic about their use in deciding between epistemic norms. Pessimism, I think, generalizes, and whatever answer one gives, whether that be a rationalist answer, a naturalist, etc., about settling other normative issues, one can give in the case of ethics.

Second, those reading this book already, I think, have some interest in our commitments on environmental issues and the questions of moral status that I take up; they are not pessimists, even if they are skeptical of the method of reflective equilibrium. For that audience, I ask whether you think that our commitments should be internally consistent. If you very strongly believe that nonsentient organisms have a welfare and find a compelling argument that there is no way to maintain that belief without also accepting that artifacts have a welfare, you are left with several choices. You might first work to undermine the argument, show it to be invalid or to have a false premise. Assuming that fails, you can either give up your initial belief about organisms or you can come to accept the conclusion about artifacts. What seems unacceptable, to me at least, is to just live with the inconsistency of recognizing the plausibility of the argument without working to change your beliefs. In order to convince you of the conclusions of this book, that is all that is really required. I'm happy for my readers to simply be convinced that they should accept the conclusions provisionally. Perhaps rationalists will complete their project to everyone's satisfaction, and that project will show the conclusions of this project to be false. But, until then, I invite you to consider the cases raised and the arguments defended, and give the reasoning a fair shake.

FOUNDATIONS OF BIOCENTRISM

1 BIOCENTRISM

The *ethic of life*, or *biocentrism*, is a view about the limits of moral concern, about which things matter from the moral point of view. In other words, it is a view about moral status.[5] According to biocentrism, every living organism, every plant, wombat, and cicada, has moral status; each is to be taken into account in our moral deliberations to at least some extent. This is the core commitment of biocentrism.

Which beings have moral status is a, perhaps *the*, central theoretical question of environmental ethics. It is at the heart of almost every question concerning our environmental obligations: at the heart of disagreements over land use, environmental restoration, hunting, agriculture, geoengineering, and assisted migration. While everyone agrees that ecosystems, nonsentient organisms, and sentient organisms matter, in some sense, from the moral point of view—that, for example, it would be wrong to destroy an ecosystem or wipe out every nonsentient organism—there is deep disagreement over which of these types of things matter fundamentally or for their own sake. The central theoretical question concerns this kind of fundamental or *direct moral status* as opposed to *indirect moral status*.

To say that a thing has moral status, generally, is to say that we are constrained, or potentially constrained, with respect to how we may act regarding that thing; a thing has moral status if we have

---

5. The term "moral status" is used differently in the various literatures in which it appears. It is common, for example, in the animal ethics literature and in the literature surrounding the human moral status, to distinguish moral status from moral standing, the former admitting of degrees or types, and the latter being a binary notion. In the environmental ethics literature, there are a host of terms that are used interchangeably with, or as a substitute for, "moral status." My primary concern is with whether nonsentient life matters *at all* or *in any way* from the moral point of view and not with *where nonsentient things stand* relative to other beings with such status/standing.

a pro tanto reason not to just do as we please regarding that thing.[6] A first gloss on the distinction between indirect and direct moral status is that a being with direct moral status is the kind of thing that moral agents ought to be *responsive to* as opposed to merely *responsive about*. The distinction is most easily gotten at via example. Consider that I have a large rock. The rock is both owned by me and on my property. Next to it is a sign indicating that this is my favorite rock and that I love the shape of it very much. One day, along comes a person who sees the rock and decides, for whatever reason, to destroy it. Whether or not this person is all things considered permitted to do so depends on the reasons for destroying the rock, but there is at least a pro tanto reason for this agent to constrain their action. The rock figures into, or should figure into, the moral deliberation of this agent in this way, and so, it has moral status. But the rock doesn't matter fundamentally. If this rock were on the agent's property or were found, unowned, somewhere, it is plausible that the agent's actions would not be so constrained. The agent's actions are constrained in this case because *I* have a kind of moral status that isn't dependent in the way the rock's is. The rock has indirect status, and I have direct moral status.

It is relatively simple to characterize indirect moral status. A thing has indirect moral status when it stands in particular relationships to things that have direct moral status. A full characterization depends on being able to identify all the particular relationships that things might stand in relative to things with direct moral status. I don't intend to offer an exhaustive list of the relevant relationships, but there are several. For example, while I believe that my dog has direct moral status, your obligations to her are not fully determined by that status. If you were to find a stray dog, it would be morally permissible for you to pet her, feed her, and perhaps take her home. You may not, arguably, do any of those things with my dog because of the relationship she bears to me; your obligations regarding my dog are partly determined by the fact that she

---

6. As I'll use the terms, "pro tanto" is not synonymous with "prima facie." As I'll use the terms, to claim that "A provides a pro tanto reason/justification for B" is to claim that A is a genuine reason / bit of justification / consideration in favor of B, in the sense that A is one of the considerations that figures into an all things considered reason, even if it is ultimately outweighed or defeated by other considerations. To claim that "A is a prima facie reason/ justification for B" is to claim that A *appears* to be a reason in favor of B, but could, on reflection, turn out to be not a reason at all. If it turns out that a consideration offers only prima facie support for some further claim, then it doesn't figure into all-things-considered justification at all (in an external/objective sense). This usage of these terms is now common, but it wasn't always. Ross (1988), for example, uses "prima facie" in the way that I will use "pro tanto."

is *my* dog. We can call this type of indirect moral status "guardianship-based indirect moral status" or "ownership-based indirect moral status."[7]

I need not own an entity in order to have a morally significant interest in how it fares. Neither you nor I own any of the populations of wild animals. Still, we, and others, have obligations not to obliterate those wild populations partly because beings with direct moral status depend on their existence for their own survival. Killing all the members of a keystone species would be morally problematic, even if those individuals did not turn out to matter for their own sake, precisely because their destruction would undermine an eco-system, which might be extremely harmful to people. We can call this form of moral status "harm-based indirect moral status."

Another form of indirect moral status, "disposition-based indirect moral status," is the status a being has in virtue of the fact that your regarding or treating it in one way or another might alter your dispositions regarding beings with direct moral status. Kant, for example, believed that nonagents lack direct moral status entirely (Kant 1963). Still, he was adamant that we had obligations to treat nonagents, including many animals, well. He worried that mistreating such beings would cause us to fail in our obligations to one another, that it would instill in us dispositions to act contrary to our duty.

Given this characterization of indirect moral status, it is easy to see, even in the absence of a clear or precise characterization of direct moral status, that there is little room for deep disagreement about which entities or types of entities have indirect moral status. Even if we assume that only humans, or even a relatively small subset of humans, have direct moral status, there will be a large number of things, of all types, that have indirect moral status. The fact of our environmental dependence generates obligations in others not to allow for the destruction of important ecosystems, which in turn generates obligations toward individual organisms, both sentient and nonsentient. This is why, despite deep disagreement over which things have direct moral status, many environmental ethicists are what are called "pragmatic holists"; they endorse the view that we should act as if we have direct obligations to ecosystems since this is the simplest way to ensure that our obligations due to those with direct moral status are met (Regan 1983, 362–63; Taylor 1989, 285–307; Norton 1994).

---

7. There might be reasons not to frame the relevant relationship between my dog and me that grounds indirect duties as one of ownership, but that is the relevant relationship in the case of traditional artifacts such as televisions, cars, and computers.

While pragmatic holism might serve as a general policy, many environmental ethicists, like many philosophers, are concerned with the true nature or objects of fundamental moral concern. It is here that there is deep disagreement about moral status. From here on, I will mostly set aside issues of indirect moral status. I will use "moral status" as a synonym for "direct moral status."

An image that will be familiar to those working in the field is one of four nested concentric circles, each circle representing the domain of entities taken to have moral status with smaller circles representing smaller, nested domains (see figure below).

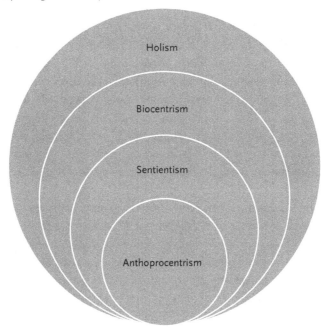

The smallest circle represents anthropocentrism, any view on which all and only humans or some subset of humans have moral status.[8] The next circle represents sentientism, the view on which all and only sentient beings have

8. This depiction is an idealization. There are lots of different ways of carving up the space. For example, anthropocentrism is used ambiguously. In some instances, it describes views on which all and only members of the species *Homo sapiens* have direct moral status. In other instances, it describes views on which all and only rational beings or moral agents have it. The anthropocentric circle could be depicted with a smaller circle inside it representing ratiocentrism and could be further carved up based on views about whether potential for rationality is sufficient for direct moral status or whether it is actual rationality, etc. Alternatively, one might think that there are rational nonhumans, and so ratiocentrism would have an odd shape, not including all that anthropocentrism includes, and including some things that it does not.

moral status. While there are debates about how sentience should be understood and exactly which capacities underwrite sentience or ground the moral status of sentient beings, there is a near consensus within environmental ethics that sentient beings have moral status.[9] In fact, this consensus might well extend to Western philosophy more generally, so long as it is understood to be just the view that all beings of sufficient cognitive complexity have direct moral status, that they matter for their own sake in some way. Consensus dissolves when we consider the next two rings. The third ring represents the ethic of life, or biocentrism, while the fourth represents what is sometimes called holism or ecocentrism.[10] Biocentrism is like anthropocentrism and sentientism in that it is *individualist*; it is only individuals, as opposed to collectives or groups, that have moral status. It differs from these views in that nonsentient individuals (organisms such as plants, mollusks, and even bacteria) have direct moral status. Holism is nonindividualist: collectives or groups of individuals themselves have moral status. There are various forms of holism distinguished by which collections or groups have such status. The relevant collections might include entire species, entire ecosystems, or the Lovelockian view that the entire biota as one has moral status (Leopold 1966; Callicott 2010; Lovelock 1988; Rolston 1989).

While the ringed image is useful for representing some aspects of fundamental disagreements over moral status, it masks two very different views about, not only which things have moral status, but the very nature of moral status itself. These approaches are the welfare approach and the intrinsic value approach. In what follows, I take up each in turn.

The Welfare Approach

The dominant approach to defending biocentrism adopts what I call the *welfare approach* to debates over moral status. The approach has three defining features. One is that moral status is characterized in terms of being morally considerable; what it means to have direct moral status is to be the type of being whose welfare is to be taken into account in moral deliberations.

---

9. While overall, I take my arguments to justify some form of sentientism, I won't defend any particular view about which cognitive capacities in particular are necessary or sufficient for moral status. I've argued elsewhere that merely being conscious (having qualia but lacking the capacity for having some form of attitude about those qualia) is not sufficient for having a welfare, and, so, moral status. See Basl 2014.

10. Varner considers both rings to be forms of biocentrism, distinguishing biocentric individualism from biocentric holism. I'll continue to use biocentrism to refer only to views that are individualist, on which only individual entities and not collectives are morally considerable.

The second concerns the argumentative strategy that is employed to defend claims that a particular entity or kind of entity is to be taken into account, a strategy of arguing for an extension of moral status to nonsentient organisms on the grounds that they share with sentient organisms that which grounds the latter's moral status, and there is no good reason to distinguish the two groups. A third feature concerns a strategy for arguing that certain entities are not to be taken into account, a strategy for excluding, say, inanimate objects, artifacts, and ecosystemic wholes from having moral status.

Moral Considerability

We can distinguish between different approaches to defending views about moral status on the basis of different conceptions or types of moral status.[11] One type or conception of moral status (call it the "moral considerability conception") is a kind of direct moral status on which having a well-being or a welfare is a necessary condition for having that kind of moral status. To have a well-being or welfare is to be the kind of thing that can be benefited or harmed, be made better or worse off; or, putting it in terms of interests, to have interests the satisfaction of which makes things go better for the thing whose interests they are and the frustration of which makes things go worse for that thing.[12]

Another feature of the moral considerability conception of moral status is that the obligation agents have toward the morally considerable is an obligation to take those interests into account in moral deliberations; the interests of the morally considerable determine, define, or ground our obligations.

We can define "moral considerability" as follows:

Moral Considerability
An entity E is morally considerable/has morally considerable interests if and only if:

---

11. I refer to this as a kind or conception of moral status in order to remain neutral about whether there are alternative species of moral status. It's possible that moral considerability is the only kind of moral status or is just identical to moral status and not a form of it.

12. Welfare, or well-being, is described differently by different participants in these debates. Taylor, for example, doesn't focus on welfare but on a thing's having a good of its own (which is different from merely having a good). Varner, and others, tend to frame the discussion in terms of having interests. Sometime interests are cashed out further in terms of represent-ability (Stone 1972; Feinberg 1980). I will, for the most part, talk of welfare, but there should be no difficulty translating the discussion to one's favored terminology. I've taken pains to avoid having my arguments rest on the particular terminology and particular connotations of that terminology.

1. E has a welfare/interests

and

2. E's welfare is morally significant.

What does it mean for welfare to be morally significant? It is not a claim about how a being is to be treated. As Singer (2009) explains in his seminal *Animal Liberation*, it is consistent with recognizing that two humans or two sentient beings are due equal consideration that we treat them differently. Given the fact that dogs, for example, have very different interests than I, it might be true that we are due equal consideration, but it is not true that the dog should have the right to vote. What it means for a thing's welfare to be morally significant, for it to be morally considerable, is that agents have a requirement to be responsive to those interests, to take them into account. What this requires is left intentionally vague since what it means to "take a being into account" depends partly on which normative theory is true. The relationship between normative theory and moral considerability is discussed in chapter 3.

A defining feature of the welfare approach to defending views about moral status is a commitment to the view that moral considerability is the *only* kind of (direct) moral status. While anthropocentrists, sentientists, and biocentrists might, and most often do, disagree about what constitutes welfare or what capacities are required for having interests, there is widespread agreement, at least between the most prominent defenders of sentientism and nearly every defender of biocentrism in the Western academic tradition, that moral considerability is not just *a* type of moral status, but the *only* type.[13]

One reason that sentientists and biocentrists might accept that moral considerability is the only game in town is that they are *welfarists*. Welfarism is, roughly, the view that the only thing of ultimate normative or axiological significance is welfare.[14] This view should be distinguished from another view, the correlate of welfarism, on which *all* welfare has fundamental normative or axiological significance. One gloss (an axiological one) is to understand

---

13. This is true of, to my knowledge, Goodpaster (1978), Taylor (1989), Attfield (1981), Varner (1998), Sterba (1998), Agar (2001), and even Stone (1972), who seems to defend an account of moral status in terms of representability, even though he seems to employ a pluralism of definitions of how a thing might have an interest or be representable. The use of "nearly" here is a philosopher's hedge. At least within debates in the environmental ethics literature, I know of no exceptions. Perhaps one might argue that Schweitzer (1969) is a paradigm example of a biocentrist who is not committed to the moral considerability conception of moral status, but even here, the description of why living things matter seems tightly connected to something like welfare even though it is expressed in terms of a will-to-live.

14. See Arneson 2002; Kraut 2009; Korsgaard 2014.

welfarism as the view that the value of a state of affairs is solely a function of welfare value. A normative gloss on welfarism is one on which it is only facts about increasing or diminishing welfare that generate reasons or constitute facts about rightness or wrongness (see, for example, Raz 1986, 194; Varner 1998, 77).[15] If welfarism is true, then whatever other properties or capacities things have, the only things that constrain agents, ultimately, are facts about welfare. If welfarism is true, the only kind of moral status that is operative is moral considerability.[16] If the correlate of welfarism is true, then having a welfare is sufficient for moral considerability.

There are those that deny welfarism. Kant, for example, didn't think that welfare was normatively fundamental, and Moore (2004) took various things to be intrinsically good independent of their being good for anyone. But one might still accept that moral considerability is the only kind of moral status. Perhaps one thinks that it is incoherent to talk about an obligation to be responsive *to* a thing that lacks a welfare, just as Kant thought it was incoherent to be responsive to something that was not an end in itself. What would it mean to be responsive to a rock? On this view, having a welfare is a necessary condition for making sense of obligation, of having reasons, etc., even if it isn't fundamentally reason-giving or valuable.

Whatever the reason biocentrists might have for thinking that moral considerability is the only game in town, it is central to their view. To my knowledge, every philosopher who thinks that welfare is not necessary for moral status, including pluralists such as Sandler (2007) who accept moral considerability as one among several types of moral status, accept that a broader range of entities that includes collectives have moral status. As we will

---

15. Welfarism need not be understood as a view about who or what bears intrinsic value. For example, welfarists need not necessarily be subject to the objection that they characterize pleasure as the only thing that is intrinsically valuable, but it is the individuals that have a welfare that are truly intrinsically valuable. The welfarist can say that individuals are the bearers of intrinsic value, but the value of a state of affairs is fully determined by how much welfare each bearer of intrinsic value has.

16. This use of operative should be distinguished from Goodpaster's (1978). Goodpaster draws a distinction between an *operative* and a *regulative* sense. A thing has moral considerability in the regulative sense if, roughly, idealized agents ought to take that thing into account. However, since we are limited beings and might be psychologically incapable of taking into account all the things that have moral considerability in the regulative sense, the set of things that agents like us actually ought take into account in our deliberations are morally considerable in the operative sense. This distinction captures the relativity of moral status: which things are morally considerable is agent-dependent. Humans can, on some plausible assumptions, have no obligations to immaterial beings with whom it is impossible for us to causally interact. But perhaps a deity does have obligations to such beings.

see, it is a commitment to moral considerability as the only game in town that purportedly justifies the exclusion of not only biological collectives, but also inanimate objects and artifacts from the domain of moral status. A cheeky way to put the point is that all the biocentrists that started working outside the welfare approach ended up giving up biocentrism. Since my aim is to undermine biocentrism, a defense of moral considerability as the sole type of moral status is unnecessary.

The Strategy of Extension

A second feature of the welfare approach to addressing issues of moral status is the way in which claims of the moral considerability of some entity or type of entity are defended, or, more accurately, the arguments for moral considerability can be modeled as instances of a particular kind of strategy. This strategy, the *strategy of extension*, is one on which a move is made from what are taken to be shared assumptions about which beings or types of beings are morally considerable, to the claim that it would be inconsistent or arbitrary not to extend moral considerability to some additional being or type of being. Here is an outline of the strategy:

1. It is assumed that some individuals—an *anchoring class*—have certain interests that are morally significant.
2. It is then argued that some other class of entities—the *target class*—have the same kind of interests.
3. Finally, it is argued that there is no morally relevant difference that would justify recognizing the moral significance of the interests had by those in anchoring class while rejecting the moral significance of the similar interests had by those in the target class.

For those that adopt the correlate of welfarism, there is an "extentionist shortcut." The third step of the strategy isn't even necessary; showing that members of the target class have interests in the same sense as members of the anchoring class is sufficient for defending the moral considerability of the target class, even if it doesn't settle questions of, for example, how to weight the various interests or trade them off against one another.[17]

---

17. This tracks the distinction that Goodpaster makes between moral considerability and moral significance (Goodpaster 1978).

As an argument for the moral considerability of some target class that can be modeled as employing the strategy, consider Singer's famous arguments for the moral status of all sentient beings (Singer 2009). Singer asserts the moral considerability of all, or almost all, human beings.[18] He also argues that at least some nonhuman animals are sentient in the sense of having an interest in pleasure or enjoyment and in the avoidance of pain and/or suffering. He then turns to the question of whether there is some difference between humans and nonhuman animals that would justify discounting or excluding from consideration sentient, nonhuman animals. He argues that there is no such morally relevant difference, or, rather, that the most typically proposed differences are not morally relevant. He does so by providing an analysis of the wrongness of racism and sexism. The problem with sexism and racism is that they allow for differential consideration of humans on the basis of morally irrelevant differences. Sex, skin color, and differences in intelligence, for example, are not morally relevant differences, differences that would justify differences in the moral considerability of humans. Furthermore, species-membership is not a morally relevant difference. Therefore, he asserts that sentient nonhumans are morally considerable.[19]

Nearly every full defense of biocentrism in the Western academic tradition can be modeled as employing the strategy of extension. To show this, it will be useful to sketch several defenses of biocentrism and show how they can be so modeled. It is not my goal to criticize these arguments here or to offer a careful analysis of their contents. It is only important that we see the ways in which these defenses are unified vis-à-vis employing the strategy of extension.

Take, for example, Paul Taylor's defense of biocentrism in *Respect for Nature* (Taylor 1989). Taylor's approach to defending biocentrism comes in two parts. He defends what he calls "The Biocentric Outlook" and "The Attitude of Respect for Nature." The biocentric outlook is a view about the natural environment, its inhabitants, and our place in it. The outlook includes a nexus of claims, for example, about our dependence on the natural environment, but also the claim that we are not metaphysically distinct or fundamentally different from other living things (Taylor 1989, chap. 3). Taylor claims that all living things are what he calls "teleological centers of life" organized toward

---

18. Singer actually argues for something stronger, not merely the consideration of animals, but equal consideration of their interests and our own.

19. More accurately, Singer places the burden of proof on those who would deny the moral considerability of sentient beings. It is always possible to propose differences between humans and other sentient beings and argue for the moral relevance of those differences. The discussion of this strategy is derived from work done in my dissertation (Basl, n.d.).

ends, and that their being so organized makes intelligible claims about what benefits and harms living things, from insects, to plants, to mammals, and what makes life go better or worse for such things. Taylor then argues that once we accept the biocentric outlook, it compels us to accept an attitude toward nature as the only consistent or coherent view to take. This attitude, the attitude of respect for nature, includes accepting that everything that has a good of its own is due respect. In other words, all living things are morally considerable (Taylor 1989, chap. 2).

It is easy to see how to model Taylor's argument using the strategy of extension. The anchoring class, for Taylor, is human beings. It is an assumption of Taylor's that human beings are due respect in virtue of our having interests or a welfare (step 1). Living things have a welfare grounded in their being teleological centers of life. It is their being organized toward ends that makes intelligible claims of the welfare of all beings, sentient and nonsentient. But, given that we are natural beings, that we are not metaphysically special, those kinds of interests are shared (step 2). Taylor does not accept the correlate of welfare as a conceptual truth (Taylor 1989, 129), but he argues that it is inconsistent with the biocentric outlook to see our welfare, but not the welfare of other organisms, as worthy of respect or consideration. To accept an asymmetry of moral considerability would be to deny the symmetry imposed on us by the biocentric outlook (step 3).

Another argument for biocentrism that can be modeled as employing the strategy of extension comes from Gary Varner's *In Nature's Interest* (1998).[20] Varner is a utilitarian, and like many utilitarians, he accepts both welfarism and the correlate. To argue for biocentrism, Varner deploys a set of cases that he takes to show that the welfare of sentient beings cannot be fully explicated by appeal to mental states, to show that that we must appeal to a notion of biologically based interests in order to make sense of the cases (Varner 1998, chap. 3). One such case involves his cat Nanci and whether it is in her interest to go outdoors. He claims, correctly, that "whether or not access to the outdoors is in the best interest of a domestic cat is an important and legitimate question" (Varner 1998, 60). Varner claims that the most plausible view about welfare, one that grounds welfare only in psychological states, cannot make sense of the importance or legitimacy of this question. On his view, our psychological interests are given by a combination of our actual and informed

---

20. Varner has since recanted his view as defended in *In Nature's Interest* partly because of criticisms that can be understood as his failing to make good on employing the strategy of extension, in particular of not being able to identify a relevant anchoring class with interests that are like those in the target class (Varner 2003, 2012).

desires. According to him, whether going outside is in the interest of Nanci is not settled by her actual desires. Furthermore, according to the theory of informed desires he adopts, Nanci is not capable of having informed desires (because she is not capable of being influenced by the relevant information, even hypothetically, that would alter her desires). So, Varner claims, the only way to accommodate these claims about Nanci's interests is to accept that she has what he calls "biological interests": interests grounded in claims about the function of her biological parts and, ultimately, in natural selection. But once we recognize the necessity of grounding some of Nanci's acknowledged interests in her biological functioning, it is a small step to recognize that nonsentient organisms have biological functions and so have biological interests. Varner employs several structurally similar cases. One involves a smoker, Maude, who has an actual and an idealized preference to continue smoking despite there being a real sense in which she has an interest in not smoking; another involves mariners in the 19th century who have an interest in vitamin C due to its effectiveness in preventing scurvy despite lacking any actual or informed desires for vitamin C.

There have been several objections to Varner's argument that Nanci's interests (and the interests of Maude and the mariners) can't be fully accommodated without appeal to biological interests (see, for example, Agar 2001, 76–77), and there are also questions and concerns, to be discussed in other chapters, about the relationship between biological function and welfare. However, notice how Varner's argument can be modeled as deploying the strategy of extension or, given Varner's welfarist commitments, the extensionist shortcut. The anchor class consists of those in the examples: Nanci, Maude, or the mariners. A case is made that they have interests of a certain kind (and are therefore morally considerable). It is then explained how it is that nonsentient organisms also have interests of that kind, that the account of welfare we use to articulate or accommodate the morally considerable interests of Nanci, Maude, and the mariners also yields the conclusion that nonsentient organisms have such interests.

As a final example, consider the defense of biocentrism in Nicholas Agar's *Life's Intrinsic Value*. On Agar's view, certain language we use is value-loaded; ascriptions or attributions of certain kinds carry with them normative significance (Agar 2001, 96). Agar takes it that our attributions of preferences or desires to other beings are loaded in this way; when we recognize other human beings as having preferences, we aren't just identifying them as having a cognitive capacity, but also attributing to them a kind of value that demands we be responsive to those preferences. In other words, ascriptions of preferences are ascriptions of moral considerability.

Agar argues that all living things can be understood as systems capable of representation, and thereby of having a type of preference (he calls it a "biopreference") (Agar 2001, chap. 6). This is possible because living things are functionally organized in a specific way such that they are teleologically organized. In virtue of being organized toward ends, they can be represented as preferring those ends. According to Agar, to the extent that we are capable of making such attributions to nonsentient beings, and, to the extent that those attributions are similar to our value-laden attributions of preferences to sentient beings, nonsentient beings are morally considerable. The fact that our preference attributions to nonsentient organisms require us to stretch our imagination a little, to develop an understanding of preference that is nonparadigmatic, is a tool that Agar uses to explain why the preferences of, for example, sentient organisms carry more weight than those of nonsentient organisms.

Again, my goal is not to raise challenges for this view, but to show that it conforms to the strategy of extension.[21] For Agar, like Varner, the anchor class includes all sentient beings to whom we are perfectly comfortable attributing moral considerability or whose moral considerability piggybacks on our attributions of preferences. An argument is then given that there is a sense in which we can attribute interests to the target class: nonsentient organisms. Agar's view is interesting because there is a sense in which there is a morally relevant difference between attributions of preferences to sentient beings and to nonsentient beings (and between various forms of nonsentient life that more closely approximate having teleologically characterized biopreferences) (Agar 2001, 97; see also chap. 6). However, it is not a morally relevant difference in the sense that it allows us to completely disregard the interests of those in the target class. Instead, it is a difference relevant to how significant those interests are when it comes to trading them off against the interests of other beings.

The Strategy of Exclusion

The strategy of extension is a useful way to understand how it is that those working within the welfare approach, including biocentrists, make the case for the moral considerability of some target class. But this is only half the story. Those working within the approach also believe that they have reasons for excluding certain entities or types of entities from the domain

---

21. For some criticisms of Agar's specific approach, see Joyce 2002; Varner 2003.

of the morally considerable; in addition to the target class, there is an *ex-clusion class*. In the case of biocentrism, even a successful application of the strategy of extension is not enough to justify biocentrism. Biocentrism is the view not only that all living things are morally considerable, but also that a range of things including ecosystems, species, biotic communities, conspecific groups, inanimate objects, and nonsentient artifacts lack moral considerability, i.e., they are in the exclusion class.[22] For the sentientist, something must be said to exclude not only these things, but also the nonsentient organisms that are the target of the strategy of extension for biocentrists.

Here, again, despite differences in the details, there is a broadly common strategy: the *strategy of exclusion*.[23] The strategy is to argue that the entities in the exclusion class do not meet the conditions for having a welfare, i.e., they are not candidates for moral considerability because they don't satisfy this necessary condition. For welfarists who adopt the correlate of welfare, that this is the strategy is obvious; questions of moral considerability just are questions of welfare. But even among those who admit a conceptual distinction between welfare and moral considerability, it has been typical to employ the strategy of exclusion as opposed to admitting that members of the exclusion class have a welfare but then identifying some morally relevant difference between members of the target and exclusion classes.

Examples of the strategy of exclusion are exhibited in all the defenses of biocentrism so far discussed. For Taylor, artifacts and wholes are not morally considerable because they do not satisfy the requirement of having a good of their own. There is a sense in which they can be benefited or harmed, but it is only in a derivative sense. Just as things can be good or bad for a heart even though a heart doesn't have a welfare (because its good is subordinate to the good of the heart-bearer), artifacts and ecosystems aren't teleological centers that have ends of their own, according to Taylor, and what is good for them simply reduces to what is good for living organisms (Taylor 1989, 7, 118–19). For Varner and Agar, biological welfare, interests, or biopreferences are

---

22. To my knowledge, every biocentrist is willing to admit that sentient artifacts, if they come to exist, will be morally considerable and are not part of the exclusion class. This just follows from the fact that biocentrists tend to recognize that cognitive capacities are an important source of interests. I'm less certain what biocentrists (or sentientists) have to say about the possibility or moral considerability of collective entities that might be modeled as having mental states or interests grounded in them.

23. Sterba (1998) is one exception to the rule. His view about how to justify exclusion will be discussed in chapter 6.

essentially tied to natural selection (Varner 1998, chap. 3; Agar 2001, chap. 5). But, they claim, artifacts and ecosystems aren't subject to natural selection and so aren't candidates for welfare.

Articulating the welfare approach not only in terms of the commitment to moral considerability as the sole type of moral status, but also in terms of the strategy of extension and the strategy of exclusion is not essential, but it is useful. Every full-throated defense of biocentrism can be understood as falling within this approach, as can the dominant approaches to defending sentientism (Singer 2009; Regan 1983), as well as some arguments for anthropocentrism (Baxter 1974). Because of this, it is easy to identify the points of contention between views within the approach; they are disagreements over which things have or lack a welfare. The burden of applying the strategy of extension is the burden of making the case that entities in the target class have a welfare, of defending a theory or account of welfare and showing that we recognize the account as grounding claims about welfare in both the anchor and the target class. The burden of applying the strategy of exclusion is showing that the account of welfare does not ground claims of welfare in members of the exclusion class. For those that endorse the correlate of welfare, that is really all there is to debates over moral status. For those who believe welfare isn't sufficient for moral status, there is a bit more arguing to do, but the resources for making the argument are, as we will see, not hard to come by.

More importantly for my purposes, that defenses of biocentrism fall under the welfare approach as characterized also makes it possible to articulate a challenge to a wide swath of views. It is because of what these views share that we can come to see that they must be rejected. The main argument of this book (the reason that biocentrism must be abandoned) can be understood as stemming from an incompatibility between the strategy of extension and the strategy of exclusion. It is my contention that there are significant constraints on what an account of the welfare of nonsentient organisms must be like. These constraints can be met, but the resulting account of welfare is one on which biocentrists will not be able to employ the strategy of exclusion to their ends. Framing the welfare approach as I have highlights the avenues available to avoid the challenge. Biocentrists might find a new strategy for defining the exclusion class, for example, but this spells the end for biocentrism; it portends the death of the ethic of life.

The Intrinsic Value Approach

The welfare approach isn't the only approach to defending claims about moral status within environmental ethics; another has its roots in Leopold's

Sand County Almanac (Leopold 1966).[24] Leopold was primarily concerned that the attitudes humans took toward the environment, even the most enlightened attitudes, were insufficient to preserve and protect the environment. He was concerned that, within the dominant approaches to regarding the environment, things either had value that was ultimately reducible to economic value or had no value at all. And so he thought we needed a sea change in our ethical outlook concerning the environment. According to Leopold, seeing ecosystemic wholes as members of the moral community, as having direct moral status, was essential to preserving the environment (Leopold 1989, 203). He paints a picture of what such an ethic would look like and tries to make the case that it is within the realm of possibility for us; that it is a natural evolution or expansion of our domain of concern (Leopold 1989, 201–3). Many environmental ethicists have been engaged in the project of trying to explicate a sense of value that wholes bear, something that can ground the direct moral status of species, ecosystems, etc., and make good on developing an environmental ethic that is adequate to meet the demands of preserving and protecting the natural environment (Callicott 1980, 2010; Rolston 1989; Throop 1999).[25]

This approach to defending claims about moral status is what I will call the *intrinsic value approach*. Unlike the welfare approach, the name here is a touch misleading. Those working within the welfare approach do not deny that there are things that have intrinsic value. On one way of talking about intrinsic value, intrinsic value is just the kind of value that grounds moral status; it is what makes a thing valuable for its own sake and what ultimately generates or grounds obligations in moral agents. Those working in the welfare approach accept that there are things with intrinsic value in this sense. They just think that intrinsic value is tightly connected to welfare, or that only bearers of welfare have intrinsic value. Those working in the intrinsic value approach think there are alternative kinds, types, or bases of intrinsic value that are not tied to welfare. For some, this motivates the view that the welfare approach is essentially anthropocentric. Since arguments for moral status are tied to the identification of an anchor class, a class that is widely taken to have moral status, we are forced to look for value by comparison to ourselves, thus missing out on other bases of value.

---

24. Another influential piece in this tradition is Routley 1973.

25. Within philosophy of ecology there has been work to show that there is a notion of an ecosystem that is more than just an illusion, or an ephemeral object with no real boundaries that couldn't possibly be an object of moral concern (Basl 2014).

Unlike the welfare approach, the intrinsic value approach can't be identified quite as neatly with some commitments about either the nature of intrinsic value or some strategy for defending views about which things have intrinsic value. The literature on intrinsic value within environmental ethics admits of myriad conceptions, and there are both value pluralists, like Sandler (2007) and Rolston (2003), who accept that moral considerability is one type of moral status among others, and intrinsic value monists, like Callicott (1990), who think that there is one form of intrinsic value, one type of moral status, but it is not a moral considerability conception.

Those working within the welfare approach have been skeptical not only about the claims that wholes have intrinsic value but also about the intrinsic value approach more generally (Varner 1998, chap. 1). For the purposes of showing that biocentrism must be rejected, I need not settle the debate between the approaches. The intrinsic value approach, whatever its merits, is not an approach available to the biocentrist. To see why this is so, first consider one of the primary motivations of the intrinsic value approach, the motivation to articulate a sense of intrinsic value that applies to wholes. It is because some philosophers think that we must go beyond biocentrism that they look to alternatives to moral considerability. Value pluralists think that there are good arguments, from within the welfare approach, that nonsentient organisms are morally considerable. For example, Sandler in defending his virtue-theoretical approach to environmental ethics takes it that all living things have a welfare and are thereby morally considerable, where this is understood as meaning that all living things fall under the scope of certain virtues, dispositions to certain kinds of responses and actions (Sandler 2007, chap. 3; see especially the section titled "Recasting Respect"). However, these same virtues can't justify our responsiveness to ecosystems or species precisely because they lack a welfare. Instead, there are different properties of these wholes that a virtuous person should be responsive to. Or, consider Rolston, who also thinks that moral considerability is a perfectly fine notion for understanding the moral status of nonsentient organisms (Rolston 2003). However, as we will see, he thinks it does not go far enough, is not inclusive enough, and so develops alternative conceptions of moral status, and the value that grounds it, to accommodate things beyond organisms.

Of course, it doesn't follow that, because those working in the intrinsic value approach are primarily motivated to justify holism, the approach couldn't ultimately be used to justify biocentrism. And there are exceptions to the general rule that those working within the intrinsic value approach are committed to holism. One of the most famous thought experiments in environmental ethics, the Last Man, is often taken as an argument for biocentrism (Sylvan 1994; Attfield 1981). In Last Man, we are to imagine a nearly empty world where the last man on Earth is deciding, or has decided, to wipe out all life that remains before he passes. We are asked to consider whether what the last man does is wrong or whether he destroys something of value. If it is wrong or he destroys something of value, it must be that life has intrinsic value.

The Last Man case might be deployed to defend biocentrism but only if we don't have similar responses to the destruction of wholes. And one of the problems with using the Last Man to test our intuitions about wholes is that is very hard to consider the destruction of an ecosystem or a species without also considering the destruction of the individuals that make them up. We could imagine that each member of the ecosystem is moved away and survives, so only an ecosystem is lost, but the wrongness of "breaking apart the ecosystem" might really be depriving the individuals of a certain kind of freedom. One of the virtues of the Last Man case and the arguments that depend on it is that it has the potential to highlight what has intrinsic value without having to identify the source of the value, but if one is to deploy it in defense of biocentrism, one must, it seems, be able to identify the source of that value, or, at the very least, to show that wholes do not also have the relevant kind of intrinsic value.

Ultimately, this means that a defense of biocentrism in the intrinsic value approach depends on a careful explanation of what besides welfare grounds or is essential for having intrinsic value. But, if we look at some of the potential sources of intrinsic value developed within the intrinsic value approach, it becomes even more clear that Biocentrists will have trouble adopting the approach as grounds for their view.

Subjective Intrinsic Value

One way of understanding intrinsic value is as "subjective intrinsic value," that is, in terms of how something is valued by valuers (Elliot 1982; O'Neill 2003; Rolston 2003; Sandler and Simons 2012). Some things are valued instrumentally, as means to an end, while others are valued noninstrumentally, as ends in themselves. I very much value having a car, but, unlike some car enthusiasts, I primarily, and maybe only, value it as a means of achieving certain ends I have, ends of getting to the grocery store or traveling with my family. On the other hand, I value my family, my friends, and my dog, not as means to an end, but in and of themselves; I want them to do well, to flourish, independent of the effects of such flourishing for me. Of course, some things are both intrinsically and instrumentally valuable. Education might be a good example, but so too are the examples of friends, family, and pets.

Trying to ground biocentrism in subjective intrinsic value seems to me impossible. First, the fact that something has value because it is valued seems to undermine claims that the intrinsic value grounds *direct* moral status. Instead moral status grounded in the valuations of valuers seems pretty clearly to be indirect. But, setting that aside, a biocentrism grounded in subjective intrinsic value faces a serious problem when it comes to exclusion. So long as the reader trusts my self-reporting, there is a proof that not only organisms, but also species and ecosystems, have subjective intrinsic value; I value, for their own sake, many species and ecosystems! Presumably, many readers need not rely on my self-reporting for such proof; you yourself probably value certain parts or aspects of nature noninstrumentally. Not only do organisms, species, and ecosystems have subjective intrinsic value; so too do many simple artifacts, such as (at least some) books, buildings, and trinkets that we are given as gifts. So long as an individual values something in a particular way, it has such value.

One way of responding to this is to revise the notion of subjective intrinsic value slightly. We might think that something has subjective intrinsic value if it is the type of thing that *all* valuers are compelled to value intrinsically. E. O. Wilson (1984) has argued that humans have a subconscious drive to connect with living beings. Perhaps we have an inherent desire or affinity for entities in the natural world, or perhaps these entities have properties which dispose us to value them intrinsically.

If it were true that we were in some sense compelled to value nature intrinsically, it would make a biocentrism grounded in subjective intrinsic value more plausible. The fact that there would be a consensus about the intrinsic value of organisms, ecosystems, or species provides grounds for discrimination; we *all* have reason to preserve and protect the environment, whereas we don't all have such reason to preserve and protect my copy of *Origin of Species*. Concerns about the directness or indirectness of the moral status of such entities, and about whether our reasons for preserving and protecting nonsentient organisms, species, and ecosystems, would also be alleviated. While the moral status of such beings would be dependent on us, and so perhaps indirect, their status and what we owe to them would be far less contingent; we couldn't simply decide that they don't matter.

However, this kind of biocentrism depends on the assumption that our biocentric leanings are widely shared. I remain skeptical that we are compelled, biologically or otherwise, to value all organisms, species, or ecosystems intrinsically. Our historical treatment of the environment and our cultural conceptions of it suggest otherwise (Cronon 1996). It may be that there is some adaptive advantage to intrinsically valuing these entities, but it isn't obvious what it is or what selection pressure would result in such valuing. And while we may be bad at introspecting some of our own mental states (see Carruthers 2010, 2011), it doesn't seem to me that I value all organisms intrinsically.[26] I recognize the instrumental value of all sorts of insects, maybe even mosquitos, and plants, maybe even poison ivy, but I don't value them for their own sake. I suspect many others, and certainly others in the tradition of the welfare approach, share these concerns. They, like me, will find it implausible that we could ground biocentrism in subjective intrinsic value. Furthermore, imagine that we can genetically engineer our children with whatever dispositions we like; we can opt to rid our children of their biophilia. Should we do so? Would we do something wrong in doing so? Would those children not somehow be mistaken in viewing all nonsentient organisms as having only instrumental value? I suspect those that feel the pull of biocentrism respond negatively to all those questions, but then there must be something more to nonsentient life than our valuing it.

Objective Intrinsic Value

Advocates within the intrinsic value approach have developed alternative accounts of the source of intrinsic value that grounds moral status that are neither subjective nor tied to welfare. These accounts attempt to identify properties of things, other than welfare, that give rise to intrinsic value in the sense of being valuable for its own sake or in and of itself. Something has *objective intrinsic value* just in case it has the relevant properties.

One example of objective intrinsic value is Rolston's (2003) *systemic value*. For Rolston, the natural environment has a kind of value in virtue of its capacity for, or its feature of, making possible things that are morally considerable. The natural environment has intrinsic value because the existence of things with another type of intrinsic value (the kind tied to welfare) depend for their existence on the natural environment. Here, in Rolston's own words:

> There are no intrinsic values, nor instrumental ones either, without the encompassing systemic creativity. It would be foolish to value the golden eggs and disvalue the goose that lays them. It would be a fundamental mistake to value the goose only instrumentally. A goose that lays golden eggs is systemically valuable. How much more so is

---

26. For further elaboration of the biophilia hypothesis and challenges, see Agar 2001, 31–40.

an ecosystem that generates myriads of species, or even, as we next see, an Earth that produces billions of species, ourselves included. (2003, 150)

The idea that, because a system generates a thing with intrinsic value, it too has such value, flirts with, if it isn't committed to, a close cousin of the genetic fallacy. We shouldn't think that value ascriptions that apply to ancestors automatically apply to offspring, or vice versa, without some argument for value transfer. But, even setting that aside, the nature of systemic value seems to generate a view about moral status that is far too inclusive to ground biocentrism. There are many things other than nonsentient individuals that are preconditions for generating things that are morally considerable.

Other potential sources of intrinsic value face similar problems. Consider what might be called *natural value*. A thing has natural value just in case it is natural. Many environmental ethicists, such as Elliot (1982), Katz (1992), and Preston (2008), have thought that naturalness is a value-adding or value-relevant property. Probably just as many have thought that naturalness is irrelevant to questions of value, or that there is no meaningful distinction between the natural and the nonnatural or unnatural. I will not rehash the debate here, primarily because it seems so obvious that naturalness value couldn't be used to justify biocentrism; whatever conception of natural you might consider, some wholes seem to count as natural.[27]

There is one property that biocentrists who wanted to defend their view from within the intrinsic value approach might appeal to as both grounding objective intrinsic value and being sufficiently exclusive: the property of being alive. On its face, this view excludes things, such as species and ecosystems, as well as artifacts. At the same time, all living things are alive! So they would all have intrinsic value. Here we have a source of objective intrinsic value worthy of considering in arguments for biocentrism!

Questions of what it means to be alive, and what properties being alive really supervene on, will be discussed in chapter 5 in the context of seeing how being alive might be relevant to having a welfare. But the short version of what is said there applies equally well here. Once we start spelling out precisely what is required to be alive, many artifacts, including things such as thermostats, will meet the conditions, and so we won't be able to exclude them

---

27. How naturalness might be employed within the welfare approach is discussed in chapter 5.

from being morally considerable or having objective intrinsic value. The alternative is to just insist that being alive doesn't reduce to properties like self-maintenance, reproductive capacity, etc., but that it is sui generis. In that case, the view that only living things have objective intrinsic value becomes very clearly question-begging. I return to these issues more fully in what follows.

2 CHALLENGES TO BIOCENTRISM

How plausible is biocentrism as developed within the welfare approach? In this chapter, I consider two kinds of challenges to defending biocentrism along these lines. The first is a general skepticism about trying to justify biocentrism independently of taking a stance about which normative theory is true. The strategy of extension, as discussed in chapter 1, makes no assumptions about which normative theory is correct, but it does presume that we can identify some anchoring class of individuals that are morally considerable. Perhaps this presumption is mistaken. Perhaps attributions of moral status are normative theory dependent. This might be fine for biocentrists who adopt the strategy; they may be very happy to commit to a normative theory. However, since my purpose is to undermine biocentrism generally, and since I wish to remain neutral on which normative theory is true, I will argue that we need take no such stance to identify an anchoring class. In showing this, it will also help to further explicate moral status.

The second sort of challenge concerns the nature of welfare and whether welfare can coherently be attributed to nonsentient organisms. As noted in the previous chapter, characterizing defenses of biocentrism in terms of the strategy of extension helps us to isolate the ways in which opponents of biocentrism might object to such arguments: they can (a) deny that there are any interests that members of the anchoring class and the target class have in common that might be morally significant, or (b) they can acknowledge those shared interests but deny their moral significance. For whatever reason, objections along the lines of (b) have been largely ignored, whereas objections along the lines of (a) are prevalent among anthropocentrists and sentientists that reject biocentrism. The basis of such objections is that welfare is inherently subjective, that bearers of welfare are essentially subjective experiencers of some kind or other.[28] According to such objections, any "interests" that plants have are derivative on the interests of sentient beings, mere illusion, anthropomorphizing run amok, or aren't attributions of welfare at all. In the second part of this chapter, I respond to these objections by first motivating objective-list views of welfare in general, and biological welfare in particular, and by defending such views against typical objections. Ultimately, what I call the *subjectivist challenge* can only be met, or only be shown to be met, by developing a theory of welfare that meets certain conditions of adequacy that the subjectivist claims a theory of nonsentient welfare cannot meet.

Moral Status and Normative Ethics

I have made little mention of the role that normative theories play in a defense of biocentrism. This might strike some readers as odd. It might seem that answers to questions of which entities in the environment we ought to be responsive to depend crucially on which normative theory is true. After all, a normative theory tells us whether an act is right. An analysis of rightness would be incomplete if it didn't tell us which entities figured into whatever constitutes rightness. Indeed, many philosophers talk and write as if there is an intimate connection between normative theory and moral status. For example, Allen Buchanan (2011, chap. 7) characterizes moral status differently depending on whether one adopts a consequentialist or contractualist perspective. According to Buchanan, consequentialists and utilitarians understand moral status as a concept which admits of degrees.[29] The capacities and properties a being has determine its relative level of moral status.[30] Contractualists, on the other hand, employ what Buchanan calls a threshold concept of moral status. On this conception, once one is judged to meet a

---

28. The possibility of denying the moral significance of recognized interests is a strategy mentioned by Cahen (2002) and by Basl and Sandler (2013a).

29. In some earlier work, Buchanan (2009) distinguishes between moral standing and moral status. The former notion is binary and the latter admits of degrees.

30. Buchanan's characterization, though perhaps coextensive, differs significantly from how utilitarians like Singer would characterize their views about moral status. Singer, for example, thinks moral status is binary; one has it or not, depending on whether one is sentient. One being's having different capacities does have different implications for how it is treated insofar as those capacities alter the range of ways, or the intensity to which, a being can suffer (Singer 2009). One way to express Singer's view is just that to have moral status means that one's suffering or enjoyment provides a pro tanto reason for action. Which action is justified requires making an all-things-considered evaluation on the basis of all the generated pro tanto reasons.

minimum threshold for certain capacities, such as the capacity to reason, one has moral status, full stop.

Fortunately, quite a lot can be said about moral status without settling debates over normative theory. To see why, notice that there are at least three distinct questions we might ask about moral status:

1. Who or what has moral status?
2. What grounds moral status?
3. How are beings with moral status to be accommodated?[31]

The first question is a question about the *bearers* of moral status, the second a question about the *source* of moral status, and the third a question about the *implications* or *significance* of moral status.

Questions about the implications of moral status are questions about how to accommodate beings with moral status. Does recognizing my moral status consist in recognizing that I have particular rights, or does it consist in recognizing that consequences for my welfare are to be taken into consideration when counting up the positive and negative consequences of an action? Questions about the implications of moral status map roughly onto what Goodpaster called "moral significance," in contrast to moral considerability; they are questions about, for example, weighing interests against one another and making trade-offs. These questions presuppose that a being has moral status, and answers to these questions quite obviously depend on answers to questions of normative theory. If hedonistic act-utilitarianism is true, it is never an implication of having moral status that one has an inviolable right. If some form of deontological theory is true, my giving consideration to your interests will mean recognizing a set of duties I have in virtue of them and acting accordingly. If virtue theory is true, my giving consideration to your interests will consist in seeing which virtues are relevant to the interests of yours that are significant and responding virtuously. And so on and so forth. Many biocentrists have been keen to develop views about how to accommodate those with moral status within particular normative frameworks. Taylor, for example, attempts this from a deontological perspective, while *biocentric consequentialists*, like Attfield (1995, 2014; see also Carter 2005; Attfield 2005), trace out the implications of moral status and tackle questions of ordering from a consequentialist perspective.

---

31. There are, of course, many other questions we might ask, including about the *structure* of moral status, i.e., whether it is graded or hierarchical.

Questions about the source of moral status are also intimately tied to normative theory. To ask about the source of moral status is to ask about an explanation for why a given being is due accommodation, or why we ought to be responsive. Two different normative theories might be coextensive with respect to which beings count as bearers of status, and yet yield very different explanations for why those beings matter. If some form of contractualism is true, some set of beings might matter because reasonable people can't reject accommodating such and such beings in such and such way. If utilitarianism is true, what reasonable people might reject only contingently matters (assuming that they'll suffer if they are forced to tolerate what they would reasonably reject). Instead, a being has status in virtue of having the capacities such that treating it in certain ways contributes to, or diminishes, the value of states of affairs. Another nice way to get at differences in the explanation for why a being is morally considerable is via a famous objection to utilitarianism. Utilitarians and deontologists agree that typical humans are morally considerable. Deontologists see typical humans as having intrinsic value, and, in virtue of this, see their welfare as salient to moral deliberations via giving rise to, or grounding, rights. Deontologists have criticized utilitarians as misplacing the object of intrinsic value. They claim that utilitarians don't recognize humans as intrinsically valuable, that they are just vessels for that which is intrinsically valuable; they are bearers of moral status because they are cups that can be filled with that which is intrinsically valuable. Here, there is deep disagreement about that which gives rise to moral considerability, about what explains why it is that typical humans have the status that they do. The "why" of a being's being owed accommodation varies depending on the source of normativity.

The above might suggest that answers to questions about the bearers of moral status also depend on matters of normative ethics. If we can't explain why a thing has moral status independently of which normative theory is true, how can we determine which things have said status? After all, their having status depends, in an important way, on the source of such status. But, just because a normative theory ultimately explains why I have moral status, it doesn't follow that my having moral status can only be *decided* by first settling on a normative theory. At least epistemically, answers to questions about the bearers of moral status are independent of, or even prior to, answers to questions of normative theory.

Below, I offer several considerations in favor of this epistemic independence. Before doing so, let me clarify this epistemic independence a bit further. There are two distinct senses of "normative theory." In one sense, a normative theory is a theory that fills out an analysis or schema of the form: "An act is right if and only if X" or "An act is right if, only if, and because X."[32] *Maximizing hedonistic act utilitarianism* is a normative theory in this *specific* sense of normative theory. It fills out the schema in a specific way: An act is right if and only if it maximizes the ratio of enjoyment to suffering. However, there is another way that "normative theory" is used. There is a sense in which Kant, Ross, Regan, and Taylor endorse the same normative theory: a Kantian or deontological one. All these philosophers endorse different normative theories in the specific sense of "normative theory" but agree about the basic structure that a normative theory should have; they agree that a normative theory explains or identifies right actions roughly in terms of abiding by duties, or at least that all right acts are coextensive with such. Call this more inclusive sense of normative theory the *structural* sense. We can distinguish between the thesis that the bearers of moral status can be determined independently of normative theory in the specific sense and the thesis that the bearers can be determined independently of normative theory in the structural sense. Call these theses "specific independence" and "structural independence," respectively. I defend both.

Structural independence is relatively easy to defend. Consider the widespread disagreement about the bearers of moral status among philosophers who are committed to the same normative theory in the structural sense. Singer and Varner are both utilitarians, but one has defended sentientism and the other biocentrism. Similar disagreements persist between deontological biocentrists. While a deontologist is perhaps more likely to think that agency is relevant to determining whether a thing is a bearer of moral status, it is simply not true that endorsing a deontological normative theory (in the structural sense) entails a particular view about the bearers of status. Kant was an anthropocentrist,[33] Regan (1983) a sentientist, and Taylor (Taylor 1989) a biocentrist, but all are deontologists in the Kantian tradition.

We can also ask whether the kinds of reasons one has for endorsing a normative theory in the structural sense are of a kind that could compel one

---

32. I include both forms of analysis to reflect differences in how some view the relationship between metaethics and normative ethics. For some, there is a fair bit of autonomy between the domains, and a view about normative ethics does not commit one to any particular view about the truth-makers of moral claims. Others think that normative views are very tightly connected to metaethical views, such that identifying something as the true normative theory tells us precisely what it is that constitutes rightness. For a discussion of the relative independence of the metaethical and the normative and related issues, see, for example, Rawls 1999; Maguire 2015; Dreier 2002; Schroeder 2015.

33. In Kant's case, it is really a form of ratio-centrism, since a requirement for direct moral status is having ends of one's own, which for Kant is closely tied to rationality.

to adopt a view about the bearers of moral status. As becomes clear when we consider attempts to "consequentialize" other normative theories (Brown 2011; Colyvan, Cox, and Steele 2010), to show that there is a form of consequentialism (or standard decision theory) that is coextensive with another normative theory one might endorse, some philosophers take it that there are reasons to endorse some form of consequentialism, come what may. Whatever these reasons are, they are obviously not reasons that decide between normative theories in the specific sense. Whatever reasons commit me to being a consequentialist, deontologist, contractualist, or virtue theorist in the structural sense, if it isn't also a determinative of a normative theory in the specific sense, then it also doesn't commit me to a view about the bearers of status. Structural independence seems certain, except in cases where my reasons for adopting a normative theory in the structural sense are the same reasons for, or otherwise compel, my adopting a theory in the specific sense.

What of specific independence? While it might be useful for all sorts of purposes to group certain theories under the same family, each member of that family is *really* a distinct normative theory. And, at least sometimes, specific normative theories entail facts about the bearers of moral status. For example, hedonistic act utilitarianism implies that nonsentient organisms are not bearers of moral status. Thus, there is a logical relationship, at least in some cases, between specific normative theories and the bearers of status. But this doesn't undermine the *epistemic* independence picked out by specific independence. What follows from the example is just that our normative theories can have entailments about which beings are bearers of moral status. For all we know, this could be because our theories are constructed to reflect facts about which beings are bearers of moral status.

In order to show that specific independence is false, we need to have some reason to think that we can't, or shouldn't, argue about the bearers of status independently of normative theory. But we can and should. As a first piece of evidence, consider that no reasonable person doubts whether the moral equality of men and women, or between humans of different races, depends on our settling which normative theory is true. Nor has anyone seen fit to argue that this lack of doubt is because it has been shown conclusively that all plausible normative theories converge on these equality results. Rather, as will be discussed below, it is the other way around. If a normative theory somehow had the implication that men and women were due differential moral consideration, that their similar interests were not to be recognized in similar ways, we would have grounds for rejecting that normative theory. The moral equality of, at least, similarly capacitied humans (a view about the bearers of status) is a settled issue prior to normative ethical debates.

Furthermore, the phenomenon of ignoring normative theory in arguing over the bearers of status extends beyond questions of the moral equality of humans. Recall that the strategy of extension starts with assumptions about who or what has moral status; it does not include claims about which normative theory is true, and neither do the arguments of those who employ the strategy. For example, consider Singer's defense of the moral equality of all sentient beings. Singer, a utilitarian, clearly appeals to utilitarianism in spelling out the implications of his equality thesis. However, Singer's defense of the moral equality of all sentient beings is not explicitly or implicitly utilitarian. Though knowledge of Singer's ultimate view often, I think, influences people to read his arguments as being utilitarian, this is not the case; he even says, "To avoid speciesism we must allow that beings who are similar in all relevant respects have a similar right to life" (Singer 2009, 19).[34] Singer's argument proceeds by asking what is wrong with racism or sexism. The problem, in each case, is that none of the differences that could potentially hold between people of different sexes or races is a morally relevant difference, a difference that would justify disregarding or discounting benefits or harms to individuals of different races or sexes. He then turns to nonhuman animals. Such animals have a welfare, and while they differ from humans in many ways, none of those differences justify discounting or disregarding their welfare. Their interests are due equal consideration with like interests of our own. Notice that this is an application of the strategy of extension. Humans are the anchoring class and sentient beings are the target class. The case for there being no morally relevant difference between individuals in the respective classes is an argument that we know those differences to be irrelevant when considering different human groups, plus a claim that species-membership is not a morally relevant category.

More germane to the current topic is the fact that, while Singer understands moral equality in terms of equal consideration of interests, his argument for moral equality leaves open the source and implications of equal status. If it turns out, as many deontologists and contractualists think, that having certain interests grounds certain rights that are not subject to violation on utilitarian grounds, then Singer's argument will imply that all sentient beings are deserving of the rights that correspond with the interests that ground particular rights in humans. If a human's interest in not suffering grounds a right not

---

34. Singer, to my recollection, has stated that he regrets this characterization for the confusion that it has caused. In my view, the confusion arises because there has been insufficient attention paid to the way in which Singer's argument is neutral with respect to normative theory.

to be harmed, that interest in a sentient being will ground that same right. This is exactly how arguments like those of Regan proceed (Regan 1983; see especially chap. 7).

This kind of neutrality on the issue of normative theory is common in arguments about the bearers of moral status. Consider, for example, Norcross, another ardent utilitarian, and his argument that it is wrong, as a typical consumer, to take advantage of the products of factory farming. Norcross (2004) starts by asking the reader to consider a thought experiment in which an individual, Fred, loses his capacity to taste chocolate. To get it back, temporarily, Fred confines and tortures puppies to extract a chemical called "cocoamone." By extracting this chemical from the puppies and ingesting it, Fred is able to enjoy the pleasures of chocolate again. Norcross presumes that we will find Fred's behavior morally objectionable. Puppies serve as Norcross's anchoring class, and the target class is factory farm animals, such as chickens, cows, and pigs. Norcross's argument continues by assessing descriptive differences that hold between Fred's treatment of puppies and a consumer's relationship to factory-farmed products, differences such as the directness with which consumers are the cause of the harm done on factory farms and the relevant causal power one has to prevent such harms. In each case, Norcross argues that these descriptive differences are morally irrelevant.

I do not intend to assess the success of Norcross's or Singer's arguments, but only to draw attention to their neutral (with respect to normative theory) character. In the case of Norcross, the language is such that there is nothing even suggestive of utilitarianism. The same can be said of both Taylor's and Varner's arguments discussed in the previous chapter. While both are committed to a particular normative theory that informs their ultimate view about the implications of moral status (and ultimately the source of such status), their justifications for the view that nonsentient organisms are bearers of moral status do not depend on these theories. Insofar as the strategy of extension can be employed to make headway on issues of moral status, specific independence is true.

In further defense of specific independence, consider the way in which our commitments about the bearers of moral status can serve as constraints on normative theory. As a model, I'll use Scanlon's version of contractualism as defended in "What We Owe to Each Other." On Scanlon's view, what we owe to one another is determined by what moral agents can reasonably reject:

An act is wrong if its performance under the circumstances would
be disallowed by any set of principles for the general regulation of

behavior that no one could reasonably reject as a basis for informed, unforced, general agreement. (Scanlon 1998, 153)

On this view, moral agents or rational deliberators have a privileged position; what they could reasonably reject is determinate of whether a proposed act is wrong. Perhaps this is why it seems so plausible that there is a dependency relationship between normative theory and moral status. On a utilitarian view, moral agents don't seem to enjoy the same privilege that they do on this contractualist picture.

This picture does not undermine specific independence. First, as Scanlon himself acknowledges, this is not a complete normative theory; there might be other sources of normativity beyond wrongness understood in terms of reasonable rejectability (Scanlon 1998, 219). As is clear from the summary above, reasonable rejectability is sufficient for a proposed act's being wrong, it is not necessary. According to Scanlon, his view is a view about *interpersonal* morality. Those rightly concerned with the moral status of nonagents, such as the significantly mentally disabled, infants, and children, need not be concerned that such individuals fall entirely outside the scope of morality.

Second, while it is true that rational deliberators enjoy a privileged position on the contractualist scheme, it is not a privilege of moral status. Being a rational deliberator, or what such a deliberator could reasonably reject, explains why a being has the moral status that it does, but doesn't imply that nondeliberators can't be bearers of status. That will depend on whether, for example, one can reasonably reject principles of action that cause significant harm to nondeliberators when such beings lack indirect moral status.

Scanlon thinks that the grounds for what one can reasonably reject are fairly restricted; one can only appeal to prudential or self-interested reasons (reasons one has in virtue of effects on one's own welfare). If Scanlon's view of what constitutes grounds for reasonable rejection is correct, then one couldn't reasonably reject causing even great harm to nondeliberators that lack indirect moral status. Such beings include those that are significantly mentally disabled, in addition to all nonhuman animals. But, if we assume, contra Scanlon, that reasonable rejectability is all there is to the normative theory, then I take it that we have grounds for rejecting this account of reasonable rejectability and, at the very least, broadening our conception of reasonableness so as to include the welfare of beings other than ourselves. In other words, our views about the bearers of moral status serve as a constraint on normative theory. If there are good cases and arguments that show that nonsentient beings are bearers of moral status, this will constrain which normative theory we have reason to adopt. The remainder of the case for specific independence will be manifest in what follows. Since my case against the moral considerability of living things will in no way depend on a particular normative theory, to the extent that the case succeeds, both forms of epistemic independence are vindicated.

The Subjectivist Challenge to Nonsentient Welfare

The main challenge to biocentrism is a challenge to the very notion that nonsentient beings can be said to have a welfare.[35] As Singer blithely puts the challenge:

> The capacity for suffering and enjoyment is a *prerequisite for having interests at all*, a condition that must be satisfied before we can speak of interests in a meaningful way. It would be nonsense to say that it was not in the interest of a stone to be kicked along the road by a schoolboy. A stone does not have interests because it cannot suffer. Nothing that we can do to it could possibly make any difference to its welfare. (Singer 2009, 8)

I say Singer's formulation of the challenge is blithe because it too quickly goes from a true claim about stones to a generalization about all nonsentient things, including nonsentient organisms. Notice how counterintuitive this latter claim is on its face. First, we refer quite often to the health or welfare of nonsentient organisms. There is nothing mysterious, strange, or awkward in talking about what improves the health of my house plants, about identifying certain conditions as good or bad for the nonsentient life in our gardens. And, it isn't only the plants in our gardens that we assess as healthy or unhealthy, of faring well or poorly, having certain conditions be good or bad for them, it is also the plants we encounter in the wild. An oak tree invaded by a parasitic mistletoe is worse off, all else equal, than one that is not so invaded.

Second, these ascriptions, both in the wild and in our homes, are strikingly nonderivative: Though it is not odd to say that "this weed killer is good for weeds," what we mean by such an utterance is that the weed killer is good precisely because it is bad for (because it harms) plants we don't desire in

---

35. A second important challenge concerns how it is that we are to accommodate the welfare of nonsentient beings, given that there are so many of them. There are various replies available to this. Several are discussed in chapter 6, but while it is important that there be some principle(s) by which we can decide between competing interests, no biocentrists in the Western academic tradition think that it is always wrong to harm anything with an interest, and so avoid forms of this challenge on which biocentrism is claimed to make it impossible to live.

our gardens or in our lawn.[36] Third, this notion of welfare is not restricted to nonsentient organisms. We refer quite easily to a notion of biological health or welfare when assessing both human and animal health. These biological health assessments, like the health assessments of nonsentient organisms, seem entirely independent of our capacity for sentience; assessment of my biological health isn't determined by its effects on my consciousness. It's entirely possible that when I die, the doctors will be nonplussed because I'm physiologically in great health.

Critics of biocentrism recognize that we talk this way, that we have these intuitions, and that we make such attributions of welfare to nonsentient organisms and sentient organisms alike. Yet they insist that our intuitions are mistaken, that attributions of welfare are groundless, incoherent, or misplaced. Why is that? The answer is that, according to Singer, Feinberg and the like, welfare is, *essentially*, subjective.

Theories of welfare can be divided into three families of views: mentalistic views, satisfaction views, and objective-list views.[37] These families differ with respect to the nature of the constituents of welfare. In the case of mentalistic views, welfare is fully constituted by the experiences of an individual; all mentalistic views are *mind-only* theories of welfare. Hedonism is a paradigm example. Whether a life goes well or poorly is solely a function of a certain set experiences that a thing has. Theories that are in the genus of satisfaction views, on the other hand, are *mind-world* views; how well a life goes is a function of the relationship between an individual's mental states and the world. The relevant mental states are typically desires or preferences, and a life goes well to the extent that these desires or preferences are satisfied. On some theories, it is solely the actual desires of an individual that matter (Heathwood 2017), while on others the relevant desires are idealized in some way (Brandt 1979; Griffin 1988; Varner 1998).[38] Finally, objective-list views are *world-only* views; whether a life goes well or poorly is entirely a function of objective facts or facts about states of affairs. To be clear, proponents of objective-list views need not deny that facts about our mental life or the relative satisfaction of our

---

36. For a nice discussion of these sorts of attributions see Attfield 1981, 42.

37. This taxonomy is used by Parfit (1986); see also Streiffer and Basl 2011; Griffin 1988.

38. There is a distinction between theories that merely restrict which desires the satisfaction of which constitute well-being and theories that truly idealize the relevant desires. Someone may endorse the view that it is only a subset of your actual desires or a certain kind of your actual desires (Heathwood 2018) which are relevant to well-being without thinking that the relevant desires are those you would have under conditions of full information, or if you were an ideal agent.

desires are irrelevant to our welfare. Most proponents of such views believe that being in the state of affairs such that one has pleasant experiences, or has certain desires satisfied, contributes to one's welfare; objective facts about how your mental life is going are *partly* constitutive of your welfare. However, many proponents of such views think that there are other constituents of welfare that are entirely divorced from our mental life. Some examples include autonomy, authenticity, having certain relationships, such as friendship, achieving certain epistemic states, or realizing one's capacities or ends (Sen 1993; Nussbaum 2013).

A theory of welfare is subjective just in case having cognitive capacities is a necessary condition for having a welfare. All mentalistic and satisfaction views are subjective in this sense. One way that sentientists can, and have, argued against the welfare of nonsentient welfare is by defending a view within these families. It is only objective-list views that make possible welfare for nonsentient organisms, and so, biocentrists are committed to such a view.[39] A first step, then, in defending biocentrism is making plausible objective-list theories.

In Favor of Objective-List Views

By evaluating some particular theories of welfare and the problems they face, a prima facie case for some form of objective-list view emerges.[40] First, consider attitudinal hedonism, a mentalistic view according to which the sole constituents of welfare are the presence or absence of enjoyment and suffering.[41] This view is subject to an objection based on one of philosophy's most famous thought experiments: the Experience Machine (Nozick 1974). I won't rehearse the entirety of the thought experiment, but

---

39. It is questionable whether objective theories of welfare and objective-list views are co-extensive. It might be that objective states of affairs are constitutive of welfare but that only beings with a mental life have such a welfare. On this concern see Sobel 1997.

40. An argument of this form also appears in Streiffer and Basl 2011; see also Griffin 1988.

41. Enjoyment and suffering are here understood as attitudes toward other experiences we might have. Attitudinal hedonism can be contrasted with sensory hedonism, on which it is physical sensations of pleasure and pain which are constitutive of welfare (Feldman 2002, 2004). On some views, pain is simply a type of sensation and suffering picks out an attitude one might have toward that sensation. On other views, pain has both a sensory and an attitudinal component; 'pain' picks out a certain class of sensations toward which we have a certain negative attitude. On this latter view, pain and suffering might be seen as identical. By focusing on attitudinal hedonism and understanding suffering in terms of taking a negative attitude toward sensations, I hope to avoid a debate over the final analysis of pain.

in broad outline we are asked to imagine that we can plug into a machine that will put us in a simulation. The simulation is such that you will experience a much higher ratio of enjoyment to suffering than you would if you did not plug in. For those of you who would be reluctant to plug in, the thought experiment suggests that welfare is constituted by more than what's in your head.

As those of you that have taught this thought experiment in your classes are aware, not everyone is reluctant to plug into such a machine, nor should we be universally reluctant to do so. Even if there are other constituents of welfare, a life that is maximally enjoyable is a hard offer to resist, especially if other constituents of welfare are elusive, or if our life otherwise contains an unbearable amount of suffering, no matter how good it might be in other purported respects. However, the thought experiment can be strengthened so as to, I think, convince almost anyone that plugging in is undesirable. Instead of an experience machine that offers you a simulation that is maximally enjoyable, just consider one that perfectly simulates the life you have except that it removes a single memory that causes you slight discomfort, perhaps an embarrassing quip you made once at a party that nobody holds against you or even remembers but that you reflect on occasionally, causing you to cringe. Or perhaps the simulation differs experientially only in that it removes one instance of you stubbing your toe from the overall character of your experience. In these modified cases, I hope you'll agree that plugging in seems a significantly worse option, not only compared to the original experience machine, but also compared to the option of continuing on in your actual life. And this problem is not limited to attitudinal hedonism. The very nature of mentalistic views is such that there is an experience machine that can offer you a better life, even if only a little better, than the one you actually have if that mentalistic view is true. I not only don't want to plug into the hedonist's machine, but into any machine of the kind.

One problem with attitudinal hedonism, and with mentalistic theories more generally, is that "we want to *do* certain things, not just to have the experience of doing them" (Nozick 1974, 43). Thought experiments that trade on this distinction between actuality and mere experience are easy to come by. We prefer to have real friends and a faithful partner rather than live an indistinguishable life where our friendships are a cruel joke and our partner is secretly promiscuous (Kagan 1997). Furthermore, I feel confident that such judgments aren't forced upon me by latent fears that the deceived could ultimately discover that their experiences were not veridical; we can easily imagine someone in such strong denial that any evidence likely to turn up to disabuse that individual of the veracity of their experiences would not be sufficient to do so, or that it would even cause them suffering to consider the possibility.

Thought experiments like those above and the arguments they serve to justify raise a substantial burden for mentalistic views. However, they leave satisfaction views untouched. If welfare is constituted by the satisfaction of our desires or preferences, and if we have preferences that our experiences be authentic, then even though we might also have a preference for having enjoyable experiences, plugging into an experience machine is always detrimental to our welfare to some degree. Furthermore, proponents of a satisfaction view can explain even the intuitions of those who would plug into a machine anytime it would increase their ratio of enjoyment to suffering over the life they would live; such individuals' sole or primary preference is for an enjoyable life.

However, these views suffer from other problems. A distinction is typically drawn between views on which it is the satisfaction of actual desires that is constitutive of welfare and views on which it is the satisfaction of desires that are in some way or other idealized that is constitutive of welfare.[42] Of these, idealized views are, to my mind, much more plausible. It is bad for five-year-old children to smoke and drink excessively even when they desire to and have not thought about or understood how these activities might impact their life or future options, and so have no actual desires that are undermined by smoking and drinking excessively. Furthermore, there are desires one might have, the satisfaction of which seems entirely irrelevant to one's welfare (Parfit 1986; Heathwood 2017). For example, say that I have a desire that there be an even number of stars in our solar system. I'm no astronomer, I do nothing to ascertain the fact of the matter, and almost certainly nobody will count these stars. If my actual desires are determinative of my welfare, then I'm either better or worse off at this moment depending on this fact about the solar system.

At the very least then, the relevant desires that are constitutive of welfare must survive some kind of idealization (or restriction) that limits which desires are constitutive of welfare and ensures that the desires I actually have

---

42. Here I'm glossing over a distinction between *restricted* desire theories and *idealized* desire theories. Some idealized desire theories don't really require that we idealize desirers, making them, for example, rational or supposing they have certain information they actually lack. Instead these theories understand welfare in terms of a limited set of the desirer's actual desires. The objection I raise applies to both and so I will often not distinguish between these views.

are refined enough. Different theories will offer different accounts of idealization. For example, Brandt offers an account of idealization on which a desire counts as idealized if it persists in the face of "cognitive psychotherapy," the presentation of pertinent information in a way that makes the implications of that information clear (Brandt 1979; see also Varner 1998).[43] Brandt was concerned to offer an account of idealization such that we could, at least in principle, use it to test whether some desire was of the right type. Alternatively, an ideal agent view is one on which a desire passes the test of idealization if it is one that perfectly rational agents, or perfectly rational versions of individuals, would have if they had full information or access to the complete science of the universe.

We need not consider every variation of idealization to raise a challenge for satisfaction views that rely upon them. The very fact that we must settle the issue of idealization suggests that objective-list views best capture what welfare is or is about. Idealized desire views face a dilemma: they either fail to appropriately discriminate against desires that aren't constitutive of welfare (like the desire that the number of stars be even) or presuppose that certain states of affairs are objectively welfare enhancing or diminishing. Another way to put the dilemma is that idealized views ultimately face the same problems as actual desire satisfaction views or collapse into objective-list views. We can ask of any idealized desire what makes that desire ideal, or why it would be rational to have that desire. One answer is that it is rational to desire X because it is instrumental in achieving Y, but answers of this form only push us to ask what makes it rational to desire Y. Not all of our desires are merely instrumentally rational. So what makes a given *ultimate desire* rational? If idealization is only meant to constrain the desires that play some role in realizing our ultimate desires, idealization does not rule out odd desires like the desire that the number of stars be even. Views that incorporate such views of idealization suffer from the same problems as actual desire views.

A view of idealization that avoids this objection is one on which it is preferable or rational not to have a desire regarding the number of stars, on which some desires are just better than others. But then what is it that compels idealized agents to not have such a preference or to have preferences that their life be authentic? A notion of rationality alone will not pick out such ends. Instead, it must be the value of those ends or the states of affairs which compel the idealized agent to desire them. But, on such a view, the fact of the idealized agent is otiose; the reason that the satisfaction of those desires

---

43. See Varner 1998, chap. 3, for a discussion of Brandt's view.

is good for the individual is that being in those states of affairs picked out or represented in the desires is objectively good for the individual!

An objective-list view can be constructed that accommodates a wide range of intuitions about welfare. By including enjoyment as a constituent of welfare among others, it can capture the fact that enjoyment really does always seem to improve welfare. But since the list might contain other constituents, we can explain how one's welfare can be improved by factors outside their enjoyment. By identifying certain objective states as valuable, it can explain why it is good for us to satisfy certain desires and not others. On its face, it allows for a much wider range of welfare bearers, including animals and even some humans, for which it makes no sense to talk of idealized preferences and which might not have the capacity for enjoyment, and nonsentient organisms. All of this gives us at least some reason, I think, to accept such a view.[44] At the very least, biocentrists need not be embarrassed that they must accept such a view.

The Hedonist's Revenge

There are two broad strategies subjectivists might employ to undermine biocentrism. One is to provide a defense of a particular view of welfare—for example a particular form of idealized desire theory or a particular form of hedonism— that is incompatible with nonsentient organisms having a welfare. In doing so, one might attempt to undermine the purported virtues of objective-list views, to explain why a particular subjectivist theory is, on balance, more plausible than an objective-list view which licenses nonsentient welfare.

Feldman in his "Pleasure and the Good Life" (2004, see also 2002) seeks to defend hedonism from the various objections that often motivate people to adopt alternatives, including objective-list views.[45] According to Feldman, something at the core of hedonism is correct; welfare runs through experiences of enjoyment and suffering; there is nothing that can make our lives go better or worse if it doesn't somehow affect our having experiences of enjoyment or suffering.

Feldman is a big-tent hedonist: he wants everyone to accept the core of hedonism but give everyone the leeway to endorse a form of hedonism that is responsive to the objections to hedonism they find compelling. To do so, he allows that how valuable a given experience of enjoyment is, or how disvaluable a given experience of suffering is, can be modified by other factors. He offers various

44. For further defenses of objective-list views see Rice 2013; Lin 2017.

45. See also Feldman 2002.

potential modifiers in response to objections. Consider the view he calls "verid-ical intrinsic attitudinal hedonism," VIAH for short. On this view, the value of a given enjoyed experience is higher when the enjoyed experience is veridical, i.e., the world is as it seems to be.[46] VIAH is developed precisely to preserve hedonism in the face of objections involving inauthentic friends, unfaithful spouses, and experience machines. Two individuals with identical mental lives can differ with respect to how their lives are going (according to VIAH) because their experiences may differ in veridicality, and so the value of experiences of enjoyment can differ in their contribution to welfare (Feldman 2002, 616).

Feldman also considers cases where someone has an odd or even an unpal-atable desire (Feldman 2004, 40; see also chap. 4, sec. 4). One example he uses is of a terrorist that takes enjoyment in blowing up children on playgrounds. In response, he proposes a form of hedonism he calls "desert-adjusted in-trinsic attitudinal hedonism," or DAIAH. On DAIAH whether an object of enjoyment is worthy of being enjoyed serves as a modifier for how valuable any enjoyment taken from it is. Since the terrorist enjoys something that is not worthy of enjoyment, the welfare value of the enjoyment taken in it is di-minished. This serves to avoid worries about enjoyment an individual might take in morally atrocious or worthless things.

Does Feldman undermine the motivations for adopting an alternative to he-donism? By my lights, there is still the question of what could make some objects of enjoyment worthy or unworthy; a question best answered by appealing to being in certain objective states of affairs as constitutive of welfare. Setting that aside, we can consider cases of beings that have a welfare but lack the capacity for enjoyment or suffering. For now, we can set aside nonsentient beings, since to assert their welfare would be to beg the question in favor of my own view that nonsentient organisms do have such a welfare. Instead, we consider other kinds of conscious beings. Peter de Marneffe (2003) rejects Feldman's view on the grounds that it forces us to the conclusion that beings without the capacity for enjoyment lack a welfare. He asks us to consider the following:

> Imagine someone—call him Spock—who does not enjoy things. He has some of the propositional attitudes we do—belief, approval, and

---

46. For Feldman the value of a life is a function solely of the *intrinsic* enjoyment one takes in a given experience. He introduces the notion of intrinsic enjoyment to avoid issues of double counting. He gives the example of someone enjoying to some degree some water's being in a pitcher, but this is enjoyed only because that water can be put in a glass, and that circumstance enjoyed only because it can then be drunk. I set aside here the details of his account of intrinsic enjoyment (Feldman 2004, chap. 2).

disapproval—but he does not have others—anger, fear, and, unfortunately for him, enjoyment. (de Marneffe 2003, 198)

Spock, if any form of attitudinal hedonism is true, cannot have a welfare. But the thought experiment seeks to establish that he can. De Marneffe continues the thought experiment:

> Spock is motivated by his normative judgments, his judgments that he ought to act in some way or that there is sufficient reason to do so. Spock aims to be a starship officer from a young age, and organizes all his endeavors around achieving it. He aims at it not because he thinks he will enjoy it, or because he enjoys the image of himself serving as an officer but simply because he sees it as a good thing to do, one which he is capable of doing. Now first officer of the Starship Enterprise, he aims to be a good one, to promote the safety of captain and crew and the ship's scientific and humanitarian mission, aims he is currently achieving. His life is marked by excellence in theoretical and practical reasoning and he has cooperative relationships with other rational beings. (de Marneffe 2003, 198–99)

What should we say about Spock's welfare? De Marneffe admits that we might not want to be Spock, but we shouldn't thereby conclude that Spock's life cannot go well or poorly (de Marneffe 2003, 199). Spock's life is very different from our own, but it seems to be going relatively well.

No modifier can be introduced to address the Spock example because Spock doesn't have the relevant type of experience that can be modified. And, as de Marneffe claims, this version of Spock is conceptually possible (de Marneffe 2003, 200).[47] Accordingly, Feldman's hedonism can, at best, be a theory of what makes a life go well for individuals that have the capacity for enjoyment rather than a general theory about what constitutes welfare. If we

---

47. According to the Damasio's (2005) "somatic marker hypothesis," affective or emotional states are closely tied up with our ability to make decisions. These affective states might be quite thin, such as having an immediate disgust response to a particular smell, but without them, according to the hypothesis, agents would not be able to engage in reasoning or decision-making at all. If this hypothesis is true, it would mean that Spock would have to be radically different than us, which is perhaps why de Marneffe is careful to make the argument depend only on Spock's conceptual possibility. In any case, these affective states implicated in our decision-making seem much thinner than the attitudes that Feldman's hedonism relies on.

aim to determine whether there is such a thing as nonsentient welfare, we shouldn't be moved by Feldman's hedonism(s) to reject this.

A final concern for Feldman's big-tent approach to defending hedonism concerns what we learn, assuming his defense succeeds. That is, even if we set aside cases like Spock and assume that there is some modified version of hedonism that avoids problems about how to identify worthy or unworthy pleasures, what follows? What follows is that for some relatively expansive set of nonhedonistic theories of welfare there is a modified hedonistic theory that is coextensive with respect to what makes an individual's life go well, a theory that can capture all the same judgments. This doesn't tell us that we should prefer hedonism to alternatives, just as the successful consequentializing of a nonconsequentialist normative theory doesn't tell us that some form of consequentialism is true. We only learn that if we accept that the core of hedonism is correct, we have a way to modify our view to accommodate concerns about it while maintaining what we already accept as a core insight about the nature of welfare. In order to be convinced to accept hedonism over the alternatives, there would have to be compelling reasons to think that welfare just is, fundamentally, about enjoyment, but the fact that we attribute welfare to nonsentients, at least, counts as a strike against this.

All said, I'm skeptical of this strategy for undermining biocentrism or nonsentient welfare. Which theory of welfare is correct, I think, is still up for debate. In my view, the fact that a theory of welfare accommodates intuitions that nonsentient organisms have welfare provides at least a prima facie, if not a pro tanto, reason to accept that theory. In any case, I take it that every particular theory of welfare is sufficiently controversial that arguments against nonsentient welfare that rely on a specific subjectivist theory of welfare shouldn't worry the biocentrist too much, so long as there is some motivated, plausible objective-list view that meets some general conditions of acceptability or adequacy (conditions to be discussed in what follows). For my part, I do believe that some form of objective-list view is true, and so my challenge won't depend on defending some particular subjectivist theory.

A More Promising Strategy for the Subjectivist

A better strategy for undermining biocentrism starts from a broad consensus about what a theory of welfare must be like, a set of conditions of adequacy for a theory's counting as a theory of welfare, and then argues that no account of welfare on which nonsentient organisms have a welfare satisfies those conditions. To see the difference in these strategies, consider an application of the first sort of strategy:

Strategy 1

1. If hedonism is true, nonsentient organisms lack a welfare.
2. Hedonism is true.
3. Therefore, nonsentient organisms lack a welfare.

Compared to an application of the second sort:

Strategy 2

1. Only things that exist can be said to have a welfare.
2. Ghosts do not exist.
3. Therefore, ghosts do not have a welfare.

The second of these doesn't depend on how we weigh and balance various thought experiments or intuitions about cases to determine whether to adopt some form of hedonism or some objective-list view. Instead, the first premise lays out a condition that all those who wish to defend an account of welfare should accept. This is the sort of challenge embodied in the argument of Singer, quoted previously, that attributions of welfare to nonsentient organisms are incoherent. So long as the conditions of adequacy for a theory of welfare are plausible, and very likely to be accepted by any biocentrist working within the welfare approach, an argument that cognitive capacities are required to satisfy those conditions of adequacy provides an insurmountable objection to biocentrism.

What are the relevant conditions of adequacy that would do the trick? There are three conditions of adequacy for a theory of welfare that everyone that is party to debates over moral considerability ought to accept, and that many sentientists claim cannot be met by theories of welfare that countenance the welfare of nonsentient organisms. These conditions are

> **Subject-relativity**: An account of welfare must be an account of what makes life/existence go well for the thing that lives it/whose existence it is.
>
> **Nonderivativeness**: What makes things go well or poorly for an entity must not reduce to what makes things go well or poorly for some other entity.

Nonarbitrariness: What makes things go well or poorly for some entity must be nonarbitrary or objectively specifiable.[48]

The requirement that an account of welfare be subject-relative serves as a constraint on the type of value that an account of welfare is an account of. In Sumner's words, "welfare assessments concern the prudential value of a life, namely how well it is going for the individual whose life it is."[49] The life of an organism might be good in many senses; it might be beautiful, it might be beneficial for me, but the relevant sense when we talk about welfare is the sense in which that life is good for the organism. This is another way of saying that such assessments are subject-relative, and this subject-relativity is one of the ways in which welfare value differs from other sorts of value. As examples of other types of value, Sumner contrasts prudential or welfare value with ethical, aesthetic, and perfectionist value. Since it will be relevant later, consider how perfectionist value differs from welfare value. A thing has value in the perfectionist sense when it is a good or excellent member of its kind. In Sumner's own words: "To say that something has this sort of value is to say that it is a good instance or specimen of its kind, or that it exemplifies the excellences characteristic of its particular nature" (1995, 772). Sumner doesn't defend any particular account of how assessments of perfectionist value are to be made. What's important is that perfectionist value is a type of value that is not subject-relative; for a thing to have high perfectionist value is not for some characteristic of that thing to be valuable to or for it.

Everyone should accept subject-relativity as a condition of adequacy on a theory of welfare. Think about what distinguishes those that endorse welfarism from those that deny it. The former claim that the only thing that is good is that which is good for something, the latter that things can be good without being good for. This dispute makes sense we because we recognize welfare as a distinctive type of value, one that is subject-relative.

---

48. There is a sense in which nonderivativeness and nonarbitrariness can be understood as specific ways of failing to satisfy subject-relativity. An account of welfare that doesn't meet the condition of nonderivativeness does satisfy subject-relativity because whatever is good is ultimately good for something or someone else. An account of welfare that doesn't meet the condition of nonarbitrariness fails subject-relativity because it is questionable whether the thing claimed to be good for something is good at all. Still, it is useful to distinguish these two specific ways of an account's failing subject-relativity because they pick out specific ways that it has been claimed that biocentric accounts of welfare have failed and conditions that biocentrists have explicitly tried to meet.

49. My characterization of the conditions doesn't reference 'life' in order to leave it open that nonliving things might have a welfare.

An account of welfare is not just an account of what constitutes goodness for an entity; in order for that "goodness for" to count as welfare, it must be irreducibly good for the entity. Exercise, my doctor tells me, is good for my heart, but what he really means is that it is good for me, that taking care of my heart is a means to taking care of me. If it turns out that every attribution of welfare to nonsentient organisms is really only good for them in some way that *reduces to* or is *constituted by* my welfare, then plants do not have a welfare.

It is possible to "satisfy" the conditions of subject-relativity and nonderivativeness by fiat. Consider that I might claim that the following is an account of the welfare of stones: something is good for a stone, just in case it promotes its being eroded over time. I can simply assert that whether or not this promotes the interests of any other being is irrelevant, and that, as the account states, this is an account of what is good *for* stones. However, this account is arbitrary or ad hoc. There is no independent motivation for the account or for claiming that the things that constitute welfare in fact do so. We can't objectively specify or identify anything that grounds claims about what things are welfare promoting or diminishing; they just seem to rest on my subjective view about what is good or bad for these particular things.

The condition of nonarbitrariness is meant to rule out accounts of welfare like this. It is difficult to give a precise definition of what exactly counts as arbitrariness, and perhaps arbitrariness comes in degrees, some degree of which is acceptable. Still, there are some marks of unacceptably arbitrary accounts. For example, imagine that the biocentrist gave an account of welfare on which what is good for each nonsentient organism was identified piecemeal with no account of what unifies those constituents. On such an account what is good for maple trees is to have broad leaves, while what is good for pine trees is to have narrow leaves, etc., but there is no answer to the question of why these things constitute welfare for these things or nothing at all to be found in common. An account of welfare is nonarbitrary to the extent that we can provide some motivation or some reason for thinking that the constituents of welfare really are constituents of welfare.

There may be other conditions of adequacy for a theory of welfare. For example, most of those working in the welfare approach would probably agree that the account of welfare must be naturalistic or consistent with our best scientific understanding of the world. I too think that a plausible account of welfare is naturalistic in this sense, but, given typical accounts of nonsentient welfare, it is not typically a point of contention between biocentrists and others that this condition is satisfied. For that reason, I set it aside as a formal condition. At any rate, the account of nonsentient welfare I ultimately defend as the one biocentrists ought to accept satisfies it any rate.

There is something like a condition of adequacy that has been floated as necessary to the project of defending claims of moral considerability: This condition, a normativity condition, requires that as part of defending an account of welfare, it must be shown that welfare is of normative significance, that it grounds responsiveness (Varner 1998, 71–74, 2003; Sterba 1998). For proponents of the correlate of welfarism, this is certainly a condition of adequacy. Something can only count as having a welfare if that welfare is morally salient.

I think it better not to see this as a condition of adequacy for a theory of welfare. Some (e.g., Taylor, Sandler, and myself) accept at least the conceptual possibility that something might have a welfare and that that thing's welfare might be morally irrelevant. In other words, some deny the correlate, and so this will not be a condition of adequacy on a theory of welfare. That is one reason to set is aside. A second reason is that one virtue of the strategy of extension is that the normativity of welfare comes, so to speak, for free. By identifying the anchoring class of things as morally considerable, we are already assuming that their interests are morally significant so long as the target class has interests similarly constituted.

There are other conditions of adequacy over which there might be broad but not universal agreement. For example, biocentrists might require that, at least for nonsentient things, being alive is a necessary condition for having a welfare. Sentientists might be happy to agree with this since for them anything that is morally considerable, trivially, satisfies this requirement, and subjectivists about welfare might agree for similar reasons. Most of these additional conditions of adequacy are seen as such because biocentrists take it that artifacts, at least nonsentient ones, do not have a welfare, that it is a *reductio* of any view that licenses genuine ascriptions of welfare to (most) artifacts. I deny this and take up these purported conditions of adequacy and claims of *reductio* in later chapters.

The Subjectivist Challenge

The subjectivist challenge to nonsentient welfare is the challenge of providing an account of the welfare of nonsentient organisms that meets the conditions of subject-relativity, nonderivativeness, and nonarbitrariness.

The most plausible way to interpret Singer's blithe argument is as an argument that accounts of nonsentient welfare aren't up to the challenge. It is the sort of challenge developed more fully in Feinberg's (1980) analysis of which kinds of beings can coherently be said to be bearers of rights. On Feinberg's view, rights are protective of interests, and so having interests is a necessary condition for being a potential bearer of rights. And while Feinberg admits that we often speak of plants interests, he says, "This is a case, however, where 'what we say' should not be taken seriously: we also say that certain kinds of paint are good or bad for the internal walls of a house, and this does not commit us to a conception of walls as beings possessed of a good or welfare of their own" (Feinberg 1980, 167).

What is it about interests such that nonsentient organisms can't have such things? According to Feinberg:

> Trees are not the sorts of beings who have their "own sakes" despite the fact that they have biological propensities. Having no conscious wants or goals of their own, trees cannot know satisfaction or frustration, pleasure or pain. Hence there is no possibility of kind or cruel treatment of trees. In these morally crucial respects, trees differ from higher species of animals. (Feinberg 1980, 168)

This suggests that claims of nonsentient welfare, or the accounts from which they derive, fail to meet the condition of subject-relativity. Feinberg's remark on paint suggests that perhaps ascriptions of welfare in these cases are not genuine because they fail to meet nonderivativeness.

However exactly we construe Fienberg's argument, it is not decisive. The quoted passage runs together two issues. The first is the relationship between having a welfare and having one's "own sake." This is the point that welfare is subject-relative value; to have a welfare is to be such that there can be value *for you* as opposed to value generally. The second issue concerns the relationship between consciousness and descriptions of how a thing might be treated, in this case cruelly or kindly. On one sense of what it means to be cruel it is a necessary condition for acting such that one cause pain. On this meaning of "cruel" it is clearly true that one can't be cruel to plants. Furthermore, the biocentrist can also accept that the capacities that make it possible to be cruel to a being are morally salient in the sense that it affects what precisely we owe that being, or the implications of that being's having moral status. Beings that vary with respect to their cognitive capacities also vary with respect to the ways they can be benefited and harmed (Singer 2009, 16). But not all reductions in welfare stem from cruel treatment; one does not treat someone cruelly, in this sense, by putting the person in an experience machine. Another sense of "cruel" is such that it means something like intentionally reducing welfare just for the sake of doing so. On this understanding of cruelty, claiming that one can't be cruel to plants begs the question against theories of nonsentient welfare. If we understand cruelty in this way, we need a further argument that we can't be cruel to plants.

Feinberg's ultimate defense of the claim that plants can't be said to have genuine interests is that the concept of interests (or welfare) presupposes consciousness:

> The reason is that an interest, however the concept is finally to be analyzed, presupposes at least rudimentary cognitive equipment. Interests are compounded of *desires* and *aims*, both of which presuppose something like *belief*, or cognitive awareness. (Feinberg 1980, 168)

This argument also trades on a well-documented ambiguity, similar to the one at the heart of Singer's quotation in the earlier part of this chapter (Regan 1983, 87; Taylor 1989, 63; Goodpaster 1978, 318–19). There are two logically distinct senses of "interest." In one sense, "interest" is equivalent to "taking an interest in" or "being interested in." In this sense of interest, cognitive capacities are essential to interests. I have an interest in my morning coffee and one reason for this is that I am interested in having the coffee. However, there is another sense of "interest" that is equivalent to "being in one's interest." These notions are distinct because we can take an interest in things which are bad for us, i.e., in things not in our interest in the second sense just discussed.

It is the latter sense of interest that seems most relevant to welfare, or, at least as far the conditions of adequacy are concerned, this is the relevant notion over which there is consensus on the conditions. While sometimes what I take an interest in is also in my interest (or even grounds something's being in my interest), when we inquire about welfare we are inquiring about what is in a thing's interests and whether there are other things in its interest besides what one takes an interest in. Furthermore, while the distinction alone doesn't show that there are any interests not dependent on consciousness, it carves out space for such interests. To assert that cognitive capacities are required to have interests in the latter sense is, again, to beg the very question at issue. A further argument is still needed to ground the necessity of consciousness for meeting the subjectivist challenge.

Sumner (1995, 1999) has also claimed that objective theories of welfare, the only sort of theory that can countenance nonsentient welfare on these grounds, cannot meet the subjectivist challenge specifically because they cannot meet the condition of subject-relativity.[50] Sumner's strategy is to mark

---

50. For a discussion and criticism of Sumner's view see Raibley 2010.

the distinction between subjective and objective theories in a way that is mutually exclusive and jointly exhaustive and then to argue that objective theories generally can't account for this essential feature of welfare.

Sumner characterizes subjective theories as those on which it is a necessary condition of something's being good for us that we take some form of proattitude toward it. His use of "attitude" is sufficiently vague so as to allow that different theories might characterize the relevant attitudes in different ways, for example as mere approvals, or as desires. Objective theories can be understood as those on which something can be good or bad for individuals independently of their proattitudes (Sumner 1995, 767–68).[51] Understood this way, all views in both the mentalistic and satisfaction families qualify as subjective theories, while those in the objective-list family qualify as objective.[52]

As Sumner notes, it is relatively easy to see how a subjective theory of value accounts for the subject-relativity of welfare value; the reason X is good for *me*, on such views, is that *I* take a particular attitude toward X. In his own words:

> No theory about the nature of welfare can be faithful to our ordinary concept unless it preserves its subject-relative or perspectival character. It is not difficult to see how a subjective theory is capable of satisfying this condition of adequacy. Whatever their internal differences, the defining feature of all subjective theories is that they make your well-being depend on your concerns: the things you care about, attach importance to, regard as mattering, and so on. What is crucial on such an account is that you are the proprietor or manager of a set of attitudes, both positive and negative, towards the conditions of your life. It is these attitudes which constitute the standpoint from which these conditions can be assessed as good or bad for you. It follows on this sort of account that a welfare subject in the merely grammatical

---

51. As Sobel (1997) notes, Sumner actually offers two logically distinct and incompatible characterizations of the subjective/objective distinction. On one characterization of the distinction, Sumner claims that objective theories are those on which welfare is entirely mind-independent. On such a view, many objective-list views, in fact most of the plausible ones, which are pluralistic, will count as subjective theories.

52. Strictly speaking, one could endorse an objective-list view on which the only objective states of affairs that are relevant to welfare are those in which individuals take particular attitudes. If one were convinced that it was objective facts that constituted welfare, but the only objective facts that mattered were objective facts about whether one's desires went satisfied or not, one would endorse a theory that was a subjectivist theory on Sumner's characterization but an objective-list view. I know of no one who holds such view or why one would be tempted by it rather than a more encompassing objective-list view that was properly objective in Sumner's sense.

sense—an individual with a distinct welfare—must also be a subject in a more robust sense—the locus of a reasonably unified and continuous mental life. Prudential value is therefore perspectival because it literally takes the point of view of the subject. Welfare is subject-relative because it is subjective. (Sumner 1995, 774)

In the quoted passage we get a condition of adequacy for any theory of welfare—subject-relativity—and an explanation of how subjective theories meet it—by making what has prudential value for me a function of *my* attitudes toward various objects. Objective theories cannot appeal to a welfare subject's concerns in order to ground the subject-relativity of that welfare. Subject-relativity must be satisfied in another way.

While Sumner believes no objective theory can meet his challenge, he doesn't offer any general reason that the challenge can't be met, that there is no possible way to connect welfare, understood objectively, to the subject who is the bearer of that welfare. Instead, he surveys various attempts, including one specifically biocentric attempt, to meet the challenge and argues that they fail to do so. In doing so, he puts the burden on proponents of nonsentient welfare to develop an account that meets the subjectivist challenge.

Since Sumner's challenge is an open one, the best way to respond is to develop an account of the welfare that meets the challenge. If the proof is in the pudding, one must make pudding. In this case, the pudding is a *teleological* account of welfare, one on which what is good for an individual is defined, at least in part, by its achieving *its* ends. In the next chapter, I turn to developing a teleological account of welfare that fully meets the subjectivist challenge.

3 THE ETIOLOGICAL ACCOUNT OF TELEOLOGICAL WELFARE

Even though attributions of welfare to nonsentient organisms may seem intuitive, are widespread, and don't seem to be derivative or arbitrary, if no account is available to vindicate those intuitions, then welfare-based biocentrism should be abandoned. To accept that nonsentient organisms have a welfare that meets the subjectivist challenge without being able to articulate how would make biocentrism simply a dogma.

However, there is a family of accounts of welfare that meet the subjectivist challenge; these accounts provide or allow for a nonarbitrary, nonderivative, subject-relative account of the welfare of nonsentient organisms. These are *teleological accounts of welfare*. A teleological account of welfare is one on which welfare of an entity is defined or grounded (or partly defined or grounded) by the ends, goals, or purposes of an entity.[53]

In this chapter, I will first articulate the general features of a teleological account of welfare, distinguishing it from other seemingly related views such as accounts of natural goodness. I will then explain how teleological accounts of welfare provide the resources to meet the subjectivist challenge, how it is that teleology helps to avoid charges of nonarbitrariness and nonderivativeness while meeting the challenge of grounding subject-relativity. In doing so, I will, at first, assume that nonsentient organisms are teleologically organized. However, this assumption requires defense; it raises questions about the source of that teleology. Is there a basis for

---

53. I wish to remain agnostic about whether the goal of an account of welfare is to offer an account of what 'welfare' means or whether the goal of an account is to provide an account of what grounds facts about welfare. For my part, I don't think an account of welfare in the relevant sense is a semantic thesis but a metaphysical one, but nothing hangs on it for my project.

teleological organization in nonsentient organisms, or will attempts to ground such teleology ultimately be derivative or arbitrary? In order to address this further challenge, I develop what I call an *etiological account of teleology*, an account of teleology that grounds the teleology of nonsentient organisms in their selection history.

Very few biocentrists have fully developed or defended an account of welfare, but those who do and many of those who merely gesture at such an account appeal, implicitly or explicitly, to the resources of the etiological account of teleology I develop. This is primarily because there is precedent for grounding teleology in the workings of natural selection. Proponents of etiological accounts of biological function have argued that function ascriptions, such as "The function of the heart is to pump blood" or "The function of the eye is vision," are essentially teleological. The function of the heart is determined by what the heart is for rather than merely what it does. Proponents of such accounts have argued that evolution by natural selection does the work of explaining what traits are for and so, ultimately, what their function is (Wright 1973; Neander 1988, 1991a, 1991b; Millikan 1989, 1999; Griffiths 1993). It is no wonder then that proponents of the welfare of nonsentient organisms have seen fit to draw on natural selection etiologies to do some work for them. In order to articulate the etiological account of teleology, I will provide some background on the debate over function ascriptions and how etiologists about function have sought to use natural selection to ground teleology. Doing so will help to distinguish etiological accounts of function from etiological accounts of teleology. I also address some of the objections to etiological accounts of function to explain how they may be resolved or avoided.

In later chapters, I will show that, despite the success of etiological accounts of teleology in grounding teleology in ways that avoid the subjectivist challenge, such accounts have consequences that force us to reject the normative significance of teleological welfare; while nonsentient organisms are subjects of welfare, that welfare is of no moral significance to agents like us. This is precisely because the biocentrist is unable to ground the welfare of organisms in the etiological account of teleology in a way that excludes biological collectives and traditional artifacts. Given these consequences, the biocentrist might be inclined to seek out another account of teleology that avoids the subjectivist challenge. In the final section of this chapter I consider some alternative theories of biological teleology and explain why they are implausible or, at least, are not of much use to the biocentrist.

Teleology and Natural Goodness

A teleological account of welfare is any account of welfare on which the welfare of an entity is at least partly defined, grounded, or constituted by the ends of that entity. Why only partly? Proponents of nonsentient welfare need some way to ground the welfare of nonsentient organisms that meets the subjectivist challenge, and, as argued below, nonsentient organisms being teleologically organized makes this possible. Indeed, it seems to me the only plausible way to make sense of nonsentient welfare in a way that meets the challenge. But it does not follow that teleology is the *only* source or grounds of welfare in all organisms. In other words, the proponent of nonsentient welfare might be committed to the realization of ends as part of an objective-list view of welfare without being committed to the realization of those ends as the only thing on the objective list.

Historically, there have been some who have been happy to define all of welfare teleologically. Perhaps the best-known instance of a theory of welfare that is fully teleological (or at least often claimed to be) is Aristotle's account of welfare, one on which the good of a thing is given by its telos or function (Aristotle 1999). Furthermore, it is possible to cast many subjective components of welfare in teleological terms. For example, it isn't much of a stretch to claim that to have a preference is to have an end, and what makes satisfying a preference good for us can be understood in terms of attaining the ends as understood in terms of preferences. However, proponents of say, hedonism (or of an objective-list view that has enjoyment on the list) might deny that getting enjoyment is good for one because it realizes an end; it is good for you because it is enjoyed and for no further reason.

There is no need to settle this issue here. Whether or not teleology is the sole source of welfare, teleology fully defines, grounds, or constitutes the welfare of organisms that lack the cognitive capacities that themselves allow for the generation of new ends; for *nonsentient organisms*, being teleologically organized is necessary and sufficient for having a welfare, or so I will argue. In what follows, I will simply offer an account of the welfare of nonsentient organisms, a teleological account, rather than an account of welfare generally.

While the teleology of nonsentient organisms will ultimately be grounded in the workings of natural selection rather than in the essential nature of those organisms, it is still tempting, and perhaps accurate, to see teleological accounts of welfare as direct descendants of Aristotelian views of welfare; yes, the source of teleology has changed, but, according to such accounts, the welfare of at least some entities is grounded in their ends. However, it is important to distinguish teleological accounts of welfare from another view that has its origins in Aristotle.

There has been a resurgent (nonhistorical) interest in virtue ethics generally and in Aristotle's virtue ethics in particular (Foot 2003; Hursthouse 1999). Virtue theories, as a family of normative theories, are distinctive because of the role that virtue plays in defining right action. Whereas most forms of consequentialism and deontology will define virtues in terms of those character traits or dispositions that are best suited to doing what is right, virtue ethics defines what is right in terms of virtuous character traits or dispositions. What makes an account of virtue ethics Aristotelian, as opposed to, say, Humean (see Slote 2001; Swanton 2015), is that the virtues are grounded in the concept of eudaemonia, or flourishing. For Aristotle, virtues were those dispositions of character that are necessary (though not sufficient) for flourishing. One of the primary differences between Aristotle's virtue ethics and those of the neo-Aristotelians concerns the concept of eudaemonia. For Aristotle, eudaemonia was a function of telos; it was one's ends that defined what it would mean to flourish. The problem with this view is that Aristotle's understanding of telos seems incompatible with a naturalistic understanding of our nature. This is partly because, at least prima facie, teleology itself seems to conflict with a Darwinian worldview. And, while I will defend the view that Darwinism allows for natural teleology, that teleology is very different from our distinctive purpose as understood by Aristotle. For Aristotle, our nature was given by what made us unique relative to other things or species of things. Aristotle was an essentialist about species. According to essentialism about species, there are some properties that make us human, and the boundary between humans and other species is fixed (Aristotle 1999).[54] On this view our ends are given by what is essential to our nature, and this kind of essentialism is, at least arguably, not compatible with Darwinism.[55]

Neo-Aristotelians, rather than give up on grounding eudaemonia in human nature, have sought ways to understand human flourishing that are compatible with our best understanding of the natural world. The best-known and most widely adopted approach, the natural goodness approach, was developed most thoroughly by Philippa Foot in *Natural Goodness* (2003). Very roughly, the view is that there is an important similarity between our

54. I'm not intending this to be a careful interpretation of Aristotle, but it is a common trope about Aristotle. It certainly is very similar to how Attfield (Attfield 1981, 42), for example, understands Aristotle.

55. For a nice discussion of different forms of essentialism and how they relate to modern biology see Okasha 2002. For a criticism of modern versions of essentialism see Barker 2010.

judgments, assertions, etc. of goodness in ethical and nonethical contexts, and that human flourishing, or the human good, can be modeled on the use of goodness in nonethical contexts.

Consider the judgment that for some particular dolphin, D, "D is a good dolphin." Foot and others in the natural goodness contingent think there are true sentences or judgments of this form. What is it that makes such judgments true? According the natural goodness approach, assessments of natural goodness are *kind-relative assessments*; what it means to be a good dolphin is to be a good instance of that kind, and what defines the relevant features of that kind will be compatible with, indeed partly determined by, the evolutionary processes that resulted in and unify the kind.

According to Foot and others working in the natural goodness approach, while assessments of organisms are not ethical assessments, they bear a structure similar to assessments of humans. In other words, when we claim that "Malala is a good person," we aren't trying to say something that is grammatically different than when we say that "D is a good dolphin." Rather, we are claiming that Malala is a good instance of her kind; we are making a kind-relative judgment, one that is informed by facts about the kinds of beings we are and how we evolved. Those in the natural goodness approach ground or understand human goodness in terms of human nature in this sense, those facts about us, our history, our way of life, etc. that define our kind and make kind-relative judgments possible.

I will not attempt to make trouble for neo-Aristotelian virtue ethics or the natural goodness approach, but there are theories of welfare defended by biocentrists that understand welfare along similar lines (Attfield 1981), and it is accounts of welfare on which welfare is understood in terms of what is species typical, Sumner argues, that fail to meet requirements of subject-relativity (Sumner 1995, 788–89). It is therefore important to distinguish natural goodness, and any accounts of welfare grounded in it, from teleological accounts of welfare and the approach to be defended in what follows.

There are two key differences between these types of accounts that are worth highlighting. First, the natural goodness approach and teleological accounts of welfare are, at least on their face, trying to give accounts of very different things: the natural goodness approach is an attempt to spell out a notion of "goodness," whereas a teleological account of welfare is an account of "goodness for." On the natural goodness approach, it is possible that one be a good member of one's kind but have a bad life in terms of well-being. Just as Aristotle recognized that it isn't enough to flourish that one have all the virtues, so too do neo-Aristotelians admit that there is more to flourishing than just being good. It is precisely because this difference is conflated that appeals to the natural goodness approach to ground claims of welfare are subject to Sumner's criticism that biocentric theories of welfare conflate perfectionist and prudential/welfare value. Second, on the teleological account of welfare developed below, assessments of well-being are not kind-relative; a nonsentient organism can, at least possibly, live a good life even if it lives a life that is nothing like and would be bad for another member of its kind, however that kind is understood.[56] As we will see, knowing facts about the species to which an organism belongs might be useful in determining what specifically is good for it. This is because knowing facts about the species might inform us about the evolutionary history that grounds its teleology, but the assessment or determination that something is good for a nonsentient organism is not otherwise comparative. This is, as explained above, in direct contrast to goodness assessments on the natural goodness approach.

A third potential difference concerns whether "goodness" in the natural goodness approach is a teleological concept, whether facts about human nature ground claims about our ends, and whether this plays an important role in neo-Aristotelian virtue theory. It is not clear that those working in the natural goodness approach have to take a stance on the question of teleology or how facts about species inform claims about their ends. To the extent that neo-Aristotelians can remain agnostic about questions of teleology, there is another difference between teleological accounts of welfare and natural goodness. To the extent that neo-Aristotelians believe that facts about species also ground claims about the ends of members of those species, they are free, I think, to adopt the account of teleology developed below.

Teleology and the Subjectivist Challenge

To meet the subjectivist challenge, we must provide an account of nonsentient welfare that meets the conditions of subject-relativity, nonderivativeness, and nonarbitrariness. In my view, teleological accounts of welfare satisfy these

---

56. I say "possibly" because given the way that organisms evolve, their ends will be similar and so will most often have their welfare promoted in the same ways. However, depending on how we prioritize nested interests, the welfare of two organisms of the same species might diverge. For example, consider that X amount of water is typically good for some plant. This is because it has the end of growth and water promotes that end. But it also has the end of survival and water promotes that end, in some cases, via promoting growth. Imagine that a person will kill that plant if it gets X amount of water, but not if it gets something slightly less than X. Let's say the plant will survive on slightly less water. It would be bad for this plant to get X amount of water even though it would be good for another plant or even this plant in different circumstances. We can make sense of this so long as we prioritize the end of survival over the (nested) end of growth.

conditions in a straightforward manner. To see this, start with the assumption, to be argued for below, that nonsentient organisms such as plants, insects, sea cucumbers, etc. are genuinely teleologically organized; these organisms have ends with parts and processes directed toward achieving these ends. The ends of such organisms can be described in various and oftentimes hierarchical ways. The end of a plant is to survive and reproduce; its parts and processes, its leaves and cellular processes, help achieve these ends by achieving the ends of absorbing sunlight and converting it to energy, and so on and so forth.

If we recognize nonsentient organisms as being teleologically organized, we can specify what is good or bad for such organisms, what is in their interests or what frustrates their interests, i.e., their welfare, in terms of this teleological organization. Here is a first pass at an analysis of the well-being of nonsentient organisms in terms of teleological organization:[57]

Teleological Welfare

Something, X, is good for nonsentient organism O if and only if it promotes one of O's ends. X is bad for O if and only if it frustrates one of O's ends.[58]

With this definition of the teleological welfare of nonsentient organisms in hand, we can see how the subjectivist challenge is met. What is good for nonsentient organisms is objectively specifiable by appeal to *its* ends. Specifying what is good or bad for a nonsentient organism doesn't require we unjustifiably anthropomorphize the plant, nor need we appeal to our subjective whims to identify its interests. We need only identify the organism's ends. What makes things go well for an organism is not a motley list; there is a unified picture of what makes things go well for all nonsentient organisms. Furthermore, it is intuitively motivated; realizing ends seems, quite intuitively, good for the thing that has them. This is not like the account of the welfare of stones described in the previous chapter.

Similarly, if we assume that an organism has its own ends, those interests are specifiable independent of the interests of others. What is good for the plant is not ultimately grounded in what is good for me or some other sentient

---

57. This is really just a first pass at an account. A full account will have to tell us how to deal with the fact that interests grounded in selection are nested, as discussed in note 58.

58. Sentient organisms will also have part of their welfare, the biological or teleological component, grounded in this way. However, it is not a full account of that welfare.

being. We can explain, for example, why weedkiller is bad for weeds even though it is good for us to use it on weeds. Weedkiller frustrates or undermines the ends of weeds. Or, put another way, if weedkiller didn't work to kill weeds, it would be good for them but bad for us because it satisfies the former's ends but not our own.

Finally, promoting an organism's ends is good *for that organism*. It is because a nonsentient thing is the subject of teleological organization that we can make sense of its subjecthood even though it has no subjective life. That is all that is required to satisfy subject-relativity.

To further illustrate, we can borrow a trick from the environmental ethicists toolkit. Let us imagine a world consisting only of a single nonsentient organism, say a maple tree. By hypothesis this maple tree is teleologically organized: it has the ends of growing tall, of growing leaves, of having sufficiently thick bark, of having a sufficient amount of chloroplasts, and it has these as ends, ultimately, because they contribute to further ends of survival and reproduction. Given these ends and that the tree is the subject of teleology, the thing which is teleologically organized, we can specify that sunlight is *good for the maple tree* because of the contribution it makes to the relevant ends, thus satisfying subject-relativity. That sunlight is good for the tree is objectively specifiable; it is good for the tree because of the contribution it makes to *its* ends. Finally, the welfare of our maple tree does not reduce to the welfare of some other organism; there is, by hypothesis, no other organism to reduce the maple's welfare to.

Sumner, though not a skeptic about biological teleology, is pessimistic about grounding subject-relativity in a nonsubjectivist theory of welfare and argues that teleological views do not satisfy the requirement of subject-relativity. According to Sumner, teleological views of welfare are really views about perfectionist value (Sumner 1995, 788); an organism's realizing its ends is an organism exemplifying that which makes it a good member of its kind. Here is his explanation of the fallacy committed by advocates of teleological theories of welfare:

> In the case of a tree the elision of prudential and perfectionist value is facilitated by the ambiguity of phrases such as "a good of one's own." "The good of x" can mean, among other things, either "the welfare of x" or "the goodness of x." Where a thing is capable of having perfectionist value we can certainly speak of its goodness or excellence. In that sense it undeniably has a good, which is its own by virtue of being grounded in its nature. But it is a fallacy to slip from saying that something can be good or bad of its kind to saying that it therefore has a welfare. It

is this fallacy which is committed by the teleological theory. (Sumner 1995, 789)

Sumner continues on to argue that it is easy to see that these two senses of "goodness" are conflated when we turn to more complex organisms, organisms that he thinks are properly called "subjects" because they have a subjective point of view. In human agents he thinks it especially obvious that "it is a contingent matter whether the possession of some particular excellence enhances well-being" (Sumner 1995, 789).

There are I think two things to say in response to Sumner's challenge to the ability of teleological theories of welfare to satisfy subject-relativity. First, while there are accounts of the "goodness" of organisms that do seem to be accounts of perfectionist value rather than prudential value, this hardly seems true in general. Sumner himself cites Robin Attfield's Aristotelian definition of the good of an organism, a definition on which "the good life for a living organism turns on the fulfillment of its nature" (Attfield 1981, 42). This view is one of those, mentioned above, on which the natural goodness approach is used as, or adapted to, build an account of welfare. Perhaps if those advocating the natural goodness approach define the welfare of organisms in terms of their being good instances of their kind, then they do conflate perfectionist and prudential or welfare value.[59] But notice that teleological welfare above makes no reference to an organism's being an exemplary member of its kind, and the account of teleology I will defend does not reduce teleology to species-typical behavior or other kind-related concepts. It may be that an organism that realizes its teleology will be an excellent member of its kind, but it might also turn out, depending on how we understand what constitutes membership in a kind or species and what determines which features of that kind a member must exemplify in order to be a good member of that kind, etc., that one can be a bad member of one's kind and still have a life high in well-being. That is so unless one understands being a good member of one's kind in terms of realizing the ends of the individuals of that kind, in which case perfectionist value is being defined in terms of prudential value. This might be a bad way to define perfectionist value, but it doesn't show that a teleological theory of welfare rests on a conflation of different types of value.

---

59. It is not at all clear that those working in the natural goodness approach commit themselves to an implausible form of essentialism or that they are even giving an account of welfare. For discussion see Odenbaugh 2017.

Second, once we recognize that not all teleological theories of welfare define well-being in terms of perfectionist value, the only ground we have to accept Sumner's claim that such accounts conflate these concepts of value is his insistence that a subjectivist theory of well-being is true, that subjectivity is a necessary condition for having a welfare. And yet this is what Sumner and other subjectivists who argue that any theory of the welfare of nonsentient organisms fails to meet the subjectivist challenge must show. The previous chapter was intended to motivate and create space for a view on which nonsentient organisms can have a welfare. To insist that subject-relativity can only be satisfied by a subjective theory of welfare is just that, an insistence.

Etiological Accounts of Teleology

To those unsympathetic with biocentrism or with claims that nonsentient organisms have a welfare, the case so far may seem to be built upon a house of cards. I've promised an account of welfare that avoids the subjectivist challenge. In order to deliver, I've asked you to assume that nonsentient organisms are teleologically organized and tried to explain how once we accept that they are so organized, we get an account of welfare that avoids the challenge! As noted above, proponents, like myself, of the view that nonsentient organisms have a welfare cannot simply defend a teleological account of welfare and call it a day; it must be shown that nonsentient organisms are genuinely teleologically organized, that there is an account of their teleology that ultimately avoids the subjectivist challenge.

It is fair, at this point, to wonder about the prospects of delivering an account of teleology that avoids the challenge. After all, and contra Aristotle, it might be natural to think that teleology, ultimately, has its source in the cognitive capacities of conscious beings. Humans bestow purposes or ends on things by mixing our actions with intentions. Now that we know that animals are capable of using tools (Sanz, Call, and Boesch 2014; Krützen et al. 2005), perhaps we should recognize that we are not the only purpose givers on Planet Earth, but we might still wonder whether nonsentient organisms have ends independent of beings with intentions. If we are to recognize all nonsentient organisms as teleologically organized, perhaps we must also recognize an intelligent designer as the source of that teleology. If this is so, it will be hard to square claims of the teleological organization of nonsentient organisms with the Darwinian picture of the world that many of us, myself included, accept.

As it happens, it is precisely because we live in a Darwinian world that we can make sense of teleology without intention. Evolution by natural selection is a mindless process but also a discriminating one; the selection for some traits relative to others is the (or a) primary driver of evolution.[60] That selection history, what the traits of an organism were selected for, can ground teleology without any need to appeal to intentions or design. This sort of account is an etiological account of teleology. This view, and its name, is inspired by *etiological accounts of function*, but, as we will see, there are important differences between these accounts. Still, it is useful to begin our discussion by going over some ground in the debate over the analysis of "function."

Etiological Theories of Function

Function ascriptions, both in biological and in artifactual contexts, are common. We talk of the function of the vacuum, of the pistons of a motor, of the heart, and the lungs. We recognize that some artifacts, parts, or traits have multiple functions, and that sometimes things might be used or do things that are not one of its functions. For example, we might use an engine to heat up food or use a bird's feather as a tickling device. We recognize that sometimes artifacts, their parts, or biological traits might malfunction. The engine might explode, as might a heart. Philosophers have long argued over how to ground these function ascriptions or these properties of function, how to analyze "function," about whether "function" is to be analyzed the same way when ascribing functions to artifacts and biological traits.

The etiological account of functions, first developed by Wright (1973), is an attempt to provide an analysis of "function" that makes sense of the distinction between function and accident. For Wright, it was important to distinguish some behaviors of a trait or object that are its function and others that are done as mere accidents. It is the function of the heart to pump blood and only an accident that it makes regular thump noises, it is the function of the nose to play some role in olfactory sensation or respiration and only an accident that it holds up glasses.[61] In order to address this distinction, Wright developed an account of function on which "function" is analyzed (at least partly) in terms of the causal consequences of a thing—a trait, an artifact, a

---

60. The relative importance of selection is a matter of debate. The so-called adaptationists think that all traits should be taken to be the result of natural selection, though this view has been famously criticized and is no longer dominant (Gould and Lewontin 1979; see also Lloyd 2015).

61. The actual function of the nose, apparently, has to do with regulating the temperature of air entering the lungs ("Barras 2016).

component of an artifact—that explain *why* that thing does what it does or is as it is. Here is Wright's analysis:

> The function of X is Z means (a) X is there because it does Z, (b) Z is a consequence (or result) of X's being there. (1973, 163)

Wright's analysis has two components. The first is the etiological component: a thing, X, is as it is because historically it had (or is expected to have) a certain effect. The second component of the analysis concerns what the thing does in the system of which it is a part. So, for example, on Wright's analysis, "It is the function of the heart to pump blood" means that the heart is there because it was pumping blood, historically, and that explains why the heart is as it is in the systems of which it is a part and why the heart pumps blood in current systems.

Wright's appeal to etiology helps explain the distinction between function and accident because the relevant etiologies provide an explanation of what a trait or part is *there for*. And the relevant etiology provides such an explanation because, on Wright's view, the relevant etiology, the etiology that explains what a trait is for, is a *selection etiology*. It is the function of the heart to pump blood and not make thump-thump noises because the heart is as it is because it is there to pump blood and not to make noises, and the reason that it is there to pump blood rather than to make noise is because the heart was selected for only one of those behaviors.

Wright's analysis has been subject to a variety of criticisms. Two of these are especially relevant to how etiological theories have developed. One problem concerns its generality (Boorse 1976; Neander 1991b). Wright thought that all functions ascriptions, whether ascriptions of function to traits or to parts of artifacts, were to be analyzed this way. In the case of artifacts, the relevant selection etiology would be a design etiology, one where it was our intentions that explained why something would perform as it does and so have that as its function. In the case of biological traits, where design is not available, natural selection would do the work of explaining what a trait is there for.

A second problem concerns the possibility of malfunction or of a trait with a function failing to ever correctly perform its function (Neander 1991a). A heart can fail to ever beat and yet still have the function of pumping blood. Wright's analysis allows for a distinction between function and accident but not between function and malfunction. A heart that never beats, on the Wright analysis, just doesn't have a function.

Historically, those concerned with function ascriptions in biological contexts have seemed unconcerned with whether function ascriptions were univocal, instead drawing on Wright's insights about the role that natural selection could play in analyzing biological functions.[62] For this reason, the best-known developments of etiological theories of function after Wright are etiological accounts of *biological* function (Neander 1988, 1991a, 1991b; Millikan 1989, 1999). In order to address the problem of malfunction, the requirement that a trait actually have any effect is simply dropped, as in, for example, Neander's analysis:

> It is the/a proper function of an item (X) of an organism (O) to do that which items of X's type did to contribute to the inclusive fitness of O's ancestors, and which caused the genotype of which X is the phenotypic expression, to be selected by natural selection. (Neander 1991a, 174)

In Neander's analysis, we see not only that this is taken purely as an analysis of biological function, but also that there is no need that an item actually perform its function in the organism that has it. Of course, in order to be subject to selection, items of that type had to have done something in that organism's ancestors. This is just a necessary requirement for there to have been selection for it.

For proponents of the etiological theory of biological function, function ascriptions to biological traits are essentially teleological. To say that the heart has the function of pumping blood is to say something about what the heart is *supposed to do*.[63] Furthermore, it is the teleological nature of function ascriptions that helps make sense of malfunction; it is because a thing fails to do what it is supposed to that it is malfunctioning. If function ascriptions are a-teleological, there is no sense to be made of malfunction, there is simply performance and nonperformance.

The etiologist about biological function would be in trouble if there were no basis for natural teleology.[64] If claims about biological function or claims in biology about function and malfunction depend in important ways on what traits are supposed to do, and there is nothing a trait is supposed to do, then all

---

62. Or representational content (Dretske 1995; see also Agar 2001, 101–7).

63. In the literature on functions, the language of normativity is used instead of the language of teleology. I've avoided this here because it introduces needless confusion. The proponents of etiological accounts of function do not intend that function ascriptions are essentially normative in the sense of providing reasons for action (or belief).

64. Wouters (2005) has argued that, at least for use in the biological sciences we do not need a teleological notion of function. Still, such a notion might be available and useful for other purposes.

function ascriptions and malfunction ascriptions are mistaken. But it is here that proponents of these theories appeal to Wright's insight that selection etiologies can ground teleology without appeal to a designer or a mind. Natural selection is a selection process; our traits are as they are *because* they conferred a fitness advantage on our ancestors. It is "selection for" that grounds claims about what a trait is "there for" or what it is "supposed to do." Darwinian mechanisms provide, according to the etiologist, sufficient resources for naturalizing teleology.

Etiological Functions, Nonarbitrariness, and Nonderivativeness

Above I defined the welfare of any nonsentient organism in terms promoting or frustrating that organism's ends. So long as an organism's ends are nonderivative and nonarbitrary, this teleological theory of welfare avoids the subjectivist challenge to biocentrism. Etiological accounts of functions are an appealing place to start looking for a solution to the problems of finding an account of naturalized teleology that is nonderivative and nonarbitrary. After all, etiologists about function face a similar problem even though their aims are different. Consider a theory of teleological welfare that draws on the etiological account of functions:

Functional Teleological Welfare

Something, X, is good for nonsentient organism O if and only if it promotes one of O's ends. X is bad for O if and only if it frustrates one of O's ends.

E is O's end if and only if O (or one of O's parts) has E as a function.

Or consider Varner's psychobiological theory of welfare on which to say that X is in A's interest means that

1. A actually desires X,
2. A would desire X if A were sufficiently informed across phases of A's life, or
3. X would fulfill some biological function of some organ or subsystem of A where X is a biological function of S in A if and only if
   (a) X is a consequence of A's having S and
   (b) A has S because achieving X was adaptive for A's ancestors. (Varner 1998, 68)[65]

---

65. Varner's initial version of the psychobiological theory just states number 3 in terms of satisfying a biological-based need (Varner 1998, 62). This version replaces satisfying a biologically based need with the analysis of exactly how such needs are to be analyzed.

According to both functional teleological welfare and Varner's psychobiological theory, welfare, at least nonsentient welfare, is just defined in terms of satisfying certain functions.

While functional teleological welfare (as well as the psychobiological theory) is ultimately unsuitable as an account of teleological welfare, notice that by reducing or analyzing an organism's ends in terms of functions as defined by the etiological account of functions, we get an account of teleology (and thereby welfare) that is nonarbitrary and nonderivative. It is nonarbitrary because the functions of traits on an etiological account are objectively definable in terms of the processes of natural selection; they don't rest on anthropomorphizing or ad hoc attributions of what is good for a thing. It is nonderivative because the processes that yield those functions don't depend on or simply reduce to our intentions, at least not typically.[66]

Furthermore, the teleology does not appear from thin air simply to serve as a ground for welfare; the etiological theory of function is borrowed from its other use in analyzing biological function ascriptions precisely because it undergirds the teleological nature of those ascriptions. Etiological accounts of function are meant to explain the teleological nature of function ascriptions. Accounts of teleological welfare piggyback on this and draw on that grounding of teleology to define the welfare of organisms.[67] This move isn't available if we analyze ends just any old way. Consider, for example, the following analyses of the ends of an organism:

Ends as Performance
E is O's end if and only if O (or one of O's parts) does E.

Setting aside the fact that it is extremely implausible that everything an organism does is its end, this account arrives at teleology out of nothing. There is no reason to think that O has any genuine ends or purposes or goals;

---

66. Chapter 6 will take up the ways in which intentions enter into natural selection processes and the issue of whether the processes of natural selection depend on our choices and in what ways.

67. The more notions that require teleology or are teleological, such as information, representation, or meaning, and that are therefore best articulated in terms of selection etiologies, the better the case that grounding welfare in etiological teleology is not arbitrary. We end up with a variety of reasons for accepting that selection etiologies ground teleology and so more reasons to think that selection etiologies can ground the teleological organization of organisms.

goal-directedness is sui generis or de novo. Appealing to etiological accounts of function, grounding teleology in what there was selection for, yields genuine, nonmysterious teleology.

From Etiological Functions to Etiological Teleology

A plausible account of teleological welfare should draw on insights of etiological accounts of function, in particular, borrowing the idea that teleology or ends can plausibly be grounded in selection processes, including Darwinian selection processes. But one need not endorse an etiological account of function to endorse a view on which selection etiologies ground teleology and thereby welfare. In other words, one can be an etiologist about teleology without being an etiologist about function, and there are some compelling reasons for divorcing these views from one another (Griffiths 1993).

There are some related problems concerning the different targets of analysis or explanation between etiological accounts of function and a potential etiological account of teleology. First, there is the fact that etiological accounts of function provide an analysis or explanation of function ascriptions as opposed to end or goal ascriptions. Even though I think organisms have ends and that those can be grounded in selection etiology, I'm not at all convinced that I think organisms themselves have functions.[68] It seems plausible that a tree has the ends of survival and reproduction, but not that the function of the tree is to survive and reproduce. That, to me, suggests that the tree's survival and reproduction are there to serve some other end. At the very least, etiologists about teleology and etiologists about function are interested in analyzing, explaining, or grounding different notions: function vs. end.

Even if it is appropriate to attribute functions to organisms or to understand "function" and "end" as synonyms in the context of attributing ends to nonsentient organisms, etiological accounts of function are accounts of the function of *traits* and not *organisms*. At the very least an etiological account of function must be adapted to a new purpose, since we are concerned primarily with the welfare of organisms and not their parts, so as to provide an account that takes as its target of analysis or explanation the function or end of the organism rather than its traits.

---

68. An exception to this is that organisms can have functions in larger systems. For example, a bee might have a function with respect to a hive.

One response to these issues is to simply define the end of an organism in terms of the function of its parts, as in functional teleological welfare above. If we accept that the ends of an organism are really just the collection of the function of its parts, we don't need any additional machinery in order to have an etiological account of teleological welfare.

The problem with functional teleological welfare is that it trades one explanatory puzzle for another. The etiological account of function can explain how it is or what it means for a trait to have an end, but why think that an organism is living a good life when all its traits are realizing their ends? Why is it that an organism has the ends of its collection of traits? Yes, according to functional teleological welfare an organism does have the ends of the collection of its traits, but why should this be so as opposed to seeing an organism as just a collection of traits each of which has its own ends and its own welfare and just happen to be co-located.

This might seem an odd explanatory task; it just seems obvious that the traits of an organism are there to serve the organism and that they don't have their own welfare. What is there to explain? But, at least according to the biocentrist, we should not view ecosystems as some unified thing with an end that is defined by the ends of the collection of its parts. What is it that explains why organisms are unified wholes, such that we can analyze their ends in terms of the function of their parts, but ecosystems are not?

Given this explanatory challenge, an etiological account of teleology must do something more than just define welfare in terms of etiological functions. It must explain why organisms, as opposed to their parts, are the primary units of teleological organization, and it must do so in order to explain why it is that organisms, as opposed to their parts, are the primary bearers of welfare. This is especially important when we recognize the possibility that promoting the ends or realizing the function of a trait might be bad for the welfare of an organism.

Another reason to distinguish etiological accounts of teleology from etiological accounts of function is just that it makes certain debates within the functions literature orthogonal to the issues of primary concern to those interested in providing an account of the welfare of nonsentient organisms. If an etiological account of teleology does not depend directly on the etiological analysis of function, then the proponent of the former account need not take a stand on, for example, whether function ascriptions are univocal across biological and artifactual contexts. There are also, unsurprisingly, long-standing debates about the correct analysis of "function" within particular contexts. In the biological context, for example, there are those that endorse analyses on which "function" is analyzed simply in terms of causal roles (Cummins

1975), in terms of goal-contribution (Boorse 1976, 1977, 2002), and in terms of certain dispositional properties (Bigelow and Pargetter 1987; see also Mitchell 1993).

There are also those that endorse *pluralism* about function ascriptions, meaning not only that function ascriptions are nonunivocal across biological and artifactual contexts, but that even within biology there is no single, correct analysis of "function," but different types of function ascriptions appropriate to different tasks (Amundson and Lauder 1994; Brandon 2013; Dussault and Bouchard 2016). By avoiding cashing out etiological teleology in function terms, one need not worry about the implications of monism or pluralism about function ascriptions for questions of welfare.

Perhaps the best example of a debate over functions that highlights the need to divorce etiological teleology from etiological function is a debate over how to understand biological health in medical or physiological, as opposed to environmental or ecological, contexts. Physicians, health professionals, and most of us certainly make assessments of health that are, at least prima facie, independent of any considerations of the attitudes or preferences of individuals whose health is being assessed.[69] The notion of a healthy patient does not in any way presuppose that a patient wants to be healthy, and, in fact, those who wish to undermine their own biological health are often seen as having a disorder.[70] Some lifestyles are judged to be unhealthy, in a biological sense, even when that lifestyle is fully endorsed and enjoyed by the person living it.

Philosophers have sought to give an account of biological or physiological health or dysfunction in terms of function (Boorse 1977; Wakefield 1992). One might expect that, for the same reason that etiological accounts of function seem a good fit in accounting for the welfare of nonsentient organisms, an etiological account of functions would be well suited to defining biological health. After all, the project of accounting for biological health in medical or physiological contexts seems just to be an extension of the project of accounting for biological welfare more generally. And proponents of defining or grounding biological health in terms of etiological functions have largely

---

69. Where biocentrists potentially differ from proponents of biological health in physiological or medical contexts is with respect to whether health/welfare is normative. Biocentrists are committed to the normativity of biological welfare, and it is unclear whether those who develop accounts of biological health intend them to be normative in the sense that. One challenge to biocentrism is that it conflates 'health', a nonnormative concept, with 'welfare', a normative one.

70. This issue will be taken up in the conclusion.

claimed as a virtue of their account the importance of teleology in understanding biological health (Wakefield 1992).

However, proponents of the most prominent alternative account of biological health, based on Boorse's goal-contribution account of function, do not deny the importance of teleology to biological health. According to goal-contribution accounts of function, the function of a thing is the role that it plays in contributing to some end of a system. Boorse offers a variety of analyses that he thinks adequate to the task. Here is one:

> X is performing the function Z in the G-ing of S at t, *means*
>
> At t, X is Z-ing and the Z-ing of X is making a causal contribution to the goal G of the goal-directed system S. (Boorse 1976, 80)

Notice that in Boorse's analysis, functions are not defined by selection processes and the function of a thing is not given by what that thing is supposed to do but only its contribution to some state of a system. Instead, things have functions within teleologically organized systems; it is the contribution of a thing to some goal of that system that defines its function.

While goal-contribution accounts of biological health are inconsistent with etiological accounts of function, they need not be inconsistent with an etiological account of teleology. If biological health is ultimately supposed to be objectively measurable and independent of the attitudes of bearers of health, proponents of goal-contribution accounts will need some account of the goals of a system; they will also need an account of teleology. The same arguments in favor of grounding the welfare of nonsentient organisms in teleology and for grounding teleology in natural selection apply here. By divorcing etiological teleology from etiological function, an etiological account of teleology provides resources to proponents of a broad range of views about what theory of function is most useful in defining biological health.

An Etiological Account of Teleological Welfare

An etiological account of teleological welfare is the conjunction of a teleological account of welfare and an etiological account of teleology. Here is one such account:

The Basic Etiological Account of Teleological Welfare
Something, X, is good for nonsentient organism O if and only if it promotes one of O's ends. X is bad for O if and only if it frustrates one of O's ends.

O has E as an end if and only if there was selection for E at the level
of individual organisms in the ancestral population of which O is a
descendant.

According to the basic etiological account of teleological welfare (the
basic account), welfare is defined in terms of ends just as in the simple teleological account of welfare developed above. However, it differs from functional teleological welfare in that the ends of an organism are not defined in
terms of the function of an organism or the function of its parts. Instead,
the ends of an organism are defined directly in terms of the organism's selection history. It also differs from functional teleological welfare in that it
references two dimensions of selection rather than just one; it defines ends
not just in terms of what there was selection for but also requires that, for
something to be an end of an organism, organisms must have been the unit
of selection.

This account has many virtues. By analyzing welfare in terms of teleology and analyzing teleology in terms of natural selection, the account
meets the subjectivist challenge by drawing on the insights of those that
have developed accounts of function intended to cope with the teleological nature of function ascriptions. However, since it does not define ends
in terms of functions, it avoids any challenges to etiological accounts of
function qua analysis of "function." Finally, by analyzing the ends of an
organisms not simply in terms of what there was selection for but also in
terms of the level at which selection occurs, this account makes clear why
it is that a particular organism is a unit of teleology rather than a mere
collection of teleologically organized parts. It is because traits are selected
at the *level of the organism*, because of the fitness advantage they confer on
that organism, that it is the organism that is the primary unit of teleological
organization, and that the ends of a trait are subservient to the ends of the
organism.[71]

The basic account is a good first pass, and I think something like it is the
best account of the welfare of nonsentient organisms. However, I'm not naive
enough to think that philosophers, even those sympathetic to my views about
welfare and the virtues of this account, will not be able to find flaws in the
analysis as it is. There are bound to be a variety of ways in which this analysis

---

71. The issue of the levels of selection and its relationship to welfare and to biocentrism will
be taken up in the next chapter.

must be reworded or tweaked to avoid the ingenious counterexamples which philosophers are capable of devising.[72]

A fully adequate etiological account of teleological welfare will fully address these concerns and be carefully worded so as to avoid counterexamples I've not considered. However, the basic account meets the primary requirements of and the main explanatory challenges to developing an account of the welfare of nonsentient organisms. Insofar as there are no general objections to an etiological account of teleological welfare, i.e., an objection that doesn't simply apply to something like the basic account as formulated, an etiological account of teleological welfare is sufficient to address the needs of the biocentrist. And, to the extent that an etiological account of teleological welfare is the only account that is sufficient to address the needs of the biocentrist, they should, indeed must, commit themselves to it.

The Problem of Instant Organisms

While I can't predict all the potential objections to an etiological account of teleological welfare, there is one objection that has been raised against such views that requires attention: the problem of instant organisms. An instant organism is an organism that springs into existence de novo but is otherwise intrinsically identical to some other organism. The problem for any etiological account of teleology is that instant organisms lack the relevant sort of causal history that gives rise to ends, and so, despite being intrinsically identical to some other organism that is teleologically organized, the instant organism has no ends.

The problem of instant organisms has been taken as a challenge to etiological accounts of teleological welfare (Varner 1998, 70; Holm 2017, 1077–78), but it is not unique to these accounts. The problem applies to etiological

---

72. There are also important questions that this analysis does not provide answers to. For example, there are many ways we might describe what any given trait has been selected for. Griffiths (1995), for example, discusses a disagreement over whether a mechanism that a frog uses to capture prey was selected for detecting flies, detecting insects, or detecting black spots. Griffiths thinks that we can truthfully say, in different contexts, that there was selection for any of this. This kind of pluralism might be extremely useful in accounts of teleological welfare. There is a sense in which an organ doing what it was selected for under one description might be bad for the organism. But since everything is ultimately selected for fitness or survival and reproduction, we can make sense of these cases. Typically, it will be good for the heart to do what it was selected for at levels of specific description, but those descriptions are ultimately subordinate to a description on which the selection was for survival or reproduction.

accounts of function as well as to etiological accounts of mental content and representation. Perhaps the best-known example of an instant organism is Davidson's Swampman, a creature that is intrinsically identical to Davidson but that arises suddenly from a chance collision of the relevant particles (Davidson 1987). The problem is that on Davidson's own theory of mental content, an etiological one on which a necessary condition for a systems being representational is that it was selected for that purpose, Swampman has no mental content whatsoever at the time he comes into existence, despite the fact that anyone could have a conversation with Swampman at that time that would be indistinguishable from a discussion with Davidson. Instant organisms have similarly been raised as a problem for etiological accounts of function. Boorse (1976), for example, asks us to consider rabbits, or rather something intrinsically identical to rabbits, that have come into existence de novo. On the etiological account of functions, the traits of these "rabbits" have no function; it can't be the function of the heart of such creatures to pump blood because, even though their heart does pump blood, it is not there *because* it pumps blood. There is no selection etiology that explains why it is there.

Defenders of etiological theories have had plenty to say to these objections, and the etiologist about teleological welfare can draw upon the resources developed in defense of these other etiological theories. Proponents of etiological theories have responded to the challenge raised by instant organisms in the same variety of ways that philosophers typically respond to thought experiments that highlight counterintuitive consequences of some view: they have denied that the intuitions elicited from hypothetical cases are relevant to the particular view they endorse (Millikan 1989), they have argued that the view is for one reason or another indispensable, and so, we must accept the counterintuitive conclusions, i.e., bite the bullet (Neander 1991a), and they have argued that the consequences of the view are much less counterintuitive than they seem (Neander 1991a), etc.

Neander has developed what is by my lights a very successful response to the problem of instant organisms as it applies to etiological theories of function. She argues as follows:

Suppose there are no lions. Then suppose that half a dozen lions pop into existence, we know not how. Having stared at them in stupefied amazement for some time, we eventually being to wonder about their wing-like protuberances on each flank. We ask ourselves whether these limbs have the proper function of flight. Do they? When we discover that the lions cannot actually fly because their "wings" are not strong enough, we are tempted to suppose that this settles the matter, until we remember that organismic structures are often incapable of performing their proper function because they are deformed, diseased, atrophied from lack of use, or because the creature is displaced form its natural habitat (the lions could perhaps fly in a lower gravitational field). On the other hand, often enough there are complex structures that have no functions, for instance, the vestigial wings of emus and the human appendix. The puzzle is where among these various categories are we to place the lions' "wings." I contend that we could not reliably place them in any category until we knew or could infer the lions' history. And if we were to somehow discover that the lions had no history, and were the result of an accidental and freak collision of atoms, they would definitely not belong in any of our familiar functional categories. They are not then dysfunctional either because of disease, deformity, lack of use, or because they are exiled from their natural environment. All of these require a past. Nor did they once have a function that they have now lost. Without history the usual biological/function norms do not apply. (Neander 1991a, 179–80)

Neander seems to me to be simply correct in her diagnosis of the intuitions about these discovered organisms. Without knowing about their history, it seems a mistake to attribute any function at all or at least any proper function, the sort of function for which failure to perform that function is malfunction. When we consider an organism that is intrinsically identical to one that already exists, it is tempting to see them as functionally identical. That the traits of instant rabbits lack a function is counterintuitive only because we see instant rabbits as rabbits. Once we recognize that they are in fact a different species altogether it is easy to see that our judgments about instant rabbits are sensitive to the sorts of facts that an etiological account of function would predict.

The success of a similar kind of response to the problem of instant organisms for the etiological account of teleological welfare depends on exactly how we understand the problem, what exactly the counterintuitive consequences are taken to be. Holm (2017), for example, takes the counterintuitive consequences of etiological theories to be consequences for which things have or lack a welfare. Here is a reconstruction of Holm's version of the problem of instant organisms:

The Problem of Instant Organisms: Welfare Version

1. If the etiological theory of teleological welfare is correct, there is a possibility of intrinsically identical organisms that differ with respect to whether they have a welfare (and the content of that welfare).
2. Two intrinsically identical organisms cannot differ with respect to whether they have a welfare (nor with respect to the content of that welfare).
3. Therefore, the etiological theory of teleological welfare is incorrect.

According to an etiological account of teleological welfare, premise 1 is indisputable so long as the account is taken to range over possible organisms as well as actual organisms (which I think it does and should). But it seems to me that premise 2 is false, or, at least, that the edge can be taken off the intuition that intrinsic duplicates are welfare duplicates.

Consider that Neander's lions seem not only to blunt the intuition that intrinsic duplicates are functional duplicates but also that intrinsic duplicates are teleological duplicates. Given Neander's thought experiment, we are not merely in a state of epistemic uncertainty about the function of the suddenly appearing lion's wings, but also about their end or purpose; it is precisely because we don't know what the wings are for that we don't know their function. The same considerations apply to organisms generally. If we don't know what ends the parts and processes of an organisms are organized toward, we can't assign it ends.

At first, this response seems only partial. Premise 2 above relies on the counterintuitiveness of intrinsic duplicates not being welfare duplicates rather than intrinsic duplicates not being teleological duplicates. But insofar as one is convinced that the welfare of nonsentient organisms must be cashed out teleologically in order to meet the subjectivist challenge, Neander's explaining away the intuitions about the function of wings in instant lions, coupled with the claim that the inability to assign a function to those wings supervenes on an inability to assign teleology to those wings, does provide a sufficient reason to reject premise 2. This is bad news for those like Holm who wish to leverage instant organisms as a reason to adopt a nonetiological account of teleological welfare.

Neander's response to the problem of instant organisms is nice because it doesn't merely tell us to "deal with" or accept some counterintuitive implication of an etiological account of function but actually goes some way to changing our intuitions. For me at least, my initial intuitions about the functions or ends of instant lions or their parts (and other instant organisms) are no longer my intuitions. This often happens when we are forced to think carefully about a good thought experiment that holds fixed the relevant variables.

That said, there is room to debunk the remaining intuitions we may have about intrinsic duplicates being welfare duplicates. To debunk an intuition is to try to show that the source of the intuition is such that we shouldn't grant it much epistemic weight. For example, evolutionary debunking arguments in metaethics attempt to show that we should not take any moral intuitions as a reason to accept that a moral judgment is an accurate judgment about some mind-independent moral fact.[73] Since our moral intuitions are likely to be influenced heavily by what promotes fitness, our intuitions are not very likely to track moral facts (construed as mind-independent facts).

There is some reason to be skeptical about the veracity of our judgment that an intrinsic duplicate is a welfare duplicate. First, we might be misled by the fact that, in fact, the closest things we have to intrinsic duplicates are welfare duplicates. The more similar organisms on Earth are, the more similar the content of their welfare, and, pardon the tongue twister, for every organism on Earth that has a welfare, something that is in almost every way similar to it almost certainly has a welfare as well.[74] Our (correct) judgments about actual organisms might be influencing our assent of the more general principle that all intrinsic duplicates are welfare duplicates. Second, even when we consider particular instant organisms, it might be that, mentally, we classify them as belonging to the same category as their duplicate. So, for example, if there is an instant maple tree, we might think that it is, in fact, a maple tree. Our judgment is that all maple trees have a welfare and so this maple tree also has a welfare. But an instant maple tree is not, according to many plausible species concepts, a maple tree at all.[75] Perhaps our intuition about particular instant organisms are misguided because we can't help but see them as members of the same kind as other organisms all of which we correctly judge to have a welfare.

I think the conjunction of (a) arguments for nonsentient welfare being teleological welfare, (b) Neander's response to challenges to etiological accounts

---

73. One way to put this is that debunking arguments expose intuitions as merely prima facie.

74. One exception to this is, perhaps, dead organisms which are similar to living organisms in many ways—though, perhaps whether it is good for an organism to be brought back to life depends on whether identity persists across that change.

75. On some views about species concepts, plausible species concepts will ensure that species occupy unique positions on the tree of life (Velasco 2008).

of function, and (c) the debunking considerations are sufficient to reject the second premise of the welfare version of the problem of instant organisms.

However, there is another way to frame the problem of instant organisms that requires a different sort of response. Consider the following argument:

The Problem of Instant Organisms: Moral Status Version

1. If the etiological account of teleological welfare is correct and nonsentient organisms are morally considerable, then there is a possibility of two intrinsically identical organisms that differ with respect to their moral status.
2. Two intrinsically identical organisms cannot differ with respect to their moral status.
3. So, either the etiological account of teleological welfare is incorrect or nonsentient organisms lack moral considerability.

According to this version of the problem, what is counterintuitive is not that intrinsic duplicates might not be welfare duplicates, but that intrinsic duplicates will not be moral status duplicates. It essentially asks, "How could two things be intrinsically identical but differ with respect to whether they are morally considerable?" This is not strictly a challenge to etiological accounts of teleological welfare. Indeed, given the response to the welfare version, this argument might be seen as a way of arguing against biocentrism while maintaining that nonsentient organisms have a welfare.

Again, the second premise of this argument is questionable. First, notice that it is ambiguous. Are we supposed to understand that intrinsic duplicates are moral status duplicates in all respects? This can't be true. The value that gives rise to indirect moral status often supervenes on extrinsic properties. The value of a Bernini sculpture is not the same as an intrinsic duplicate that is a forgery;[76] a wedding ring has an indirect moral status that many other intrinsic duplicates, or near enough, lack. So there are some types of moral status for which intrinsic duplicates are not moral status duplicates.

If we restrict the second premise to direct moral status, it is still ambiguous. It might mean that intrinsic duplicates are direct moral status duplicates in all respects. This too is false if all respects includes what is owed to these duplicates or how different individuals are to be taken into account in our moral deliberations. Swampman is not due the same treatment or consideration by members of Davidson's family that Davidson is due. If there is any form of justified partiality on the basis of relationships, or if, for example,

---

76. This example is adapted from Elliot (1982).

promise-making obligates the promisor to the promisee, then intrinsic duplicates do not have the same status in this sense.

If the second premise is not to be obviously false on these grounds, it must be taken to be the claim that intrinsic duplicates are duplicates with respect to whether they are bearers of direct moral status. Admittedly, this is prima facie intuitive. After all, shouldn't my being a bearer of moral status be a function solely of my intrinsic properties and not relations I bear to others? Isn't that what it means to have direct moral status? If that's so, then intrinsic duplicates should be duplicates with respect to being bearers of direct moral status.

While this argument is initially plausible, there is room to offer a debunking response. In the case of sentient organisms, it is true, I think, that intrinsic duplicates will be moral status duplicates in this restricted sense. There would be something seriously wrong with thinking that, for example, a cat was a bearer of direct moral status but that an intrinsic duplicate was not. However, in the case of sentient organisms, those committed to an objective-list view of welfare can view cats as having a welfare derived from multiple sources. The cat is a bearer of direct moral status in virtue of having a welfare that is worthy of our concern, but that welfare is multifaceted, and its moral status may arise in virtue of, or be constituted by, some interests and not others. When we turn to nonsentient organisms, there is only a single source of welfare, their teleology. It is easy to see how we might go from the correct judgment that intrinsic duplicates of sentient organisms are welfare duplicates to the judgment that all intrinsic duplicates are welfare duplicates. But the mistake we make is in failing to recognize that being sentient, unlike being teleologically organized, is a purely intrinsic property.

Ultimately, I wish to defend the view that nonsentient organisms lack moral status. I do not think this is because we can assert confidently that intrinsic duplicates are moral status duplicates. In a later chapter, I have a bit more to say about the role of intuitions about abstract principles and how they should figure into the debate over biocentrism. However, to the extent that the debunking considerations above fail to cast doubt on our intuitions about intrinsic duplicates and their moral status, this will provide an additional argument that an etiological account of teleological welfare survives but that a biocentrism that is committed to such an account does not.

Etiological Teleology and Biocentrism

What relationship does an etiological account of teleological welfare bear to biocentrism? Why should biocentrists commit to or see themselves as committed to such views in defending biocentrism? One answer is that many biocentrists simply are so committed, at least to some parts of the view. While some proponents of biocentrism in the welfare approach simply rely on the intuitive judgment that nonsentient organisms have a welfare, those that have seen fit to develop an account of welfare typically appeal, implicitly or explicitly, to some of the resources developed above. They appeal to teleology without fully developing an account of or explaining its source (Goodpaster 1978; Taylor 1989),[77] appeal to natural selection without explicitly cashing out welfare in terms of teleology (Stone 1972), or explicitly appeal to teleological notions grounded in natural selection (Varner 1998; Agar 2001).[78]

Appeals to etiological teleology also appear explicitly in criticisms of holism (Cahen 2002; Odenbaugh 2017). A common and, as we will see, mostly plausible view about ecosystems is that they are not subject to natural selection. Cahen (2002, 117) has argued that being teleologically organized is necessary (he doesn't commit to its being sufficient) for something to have a good of its own. But since ecosystems and, for Cahen, all other units except individual organisms are not subject to selection, they cannot be teleologically organized. Thus, holism is false. Similarly, Odenbaugh (2010) has argued that while ecosystems may be functionally organized, they are not functionally organized in a way that lets us make sense of their having a welfare. According to Odenbaugh, it is not plausible to claim that ecosystems (or their parts) have functions in an etiological sense. Instead, when ecologists talk about ecosystem functions, the plausible construal is that functions should be understood in terms of something like causal contributions or "Cummins functions" (Odenbaugh 2010). He argues that precisely because ecosystems don't have etiological functions, their functional organization cannot be used as a basis for ascriptions of health in a welfare sense.[79]

The fact that some biocentrists (and others who are not unfriendly to biocentrism) have seen an important connection between welfare and teleology and between teleology and etiology does not show that these connections are

---

77. Taylor (1989), as noted, also appeals to natural selection to distinguish organisms from artifacts.

78. By my lights, Varner's defense of biocentrism is the defense that most explicitly attempts to develop an account of the welfare of nonsentient organisms that meets the subjectivist challenge by appeal to an etiological account of teleological welfare, though Agar's defense is closely aligned. As noted above, Varner defines biological interests explicitly in terms of the etiological account of function.

79. Odenbaugh talks of norms of performance rather than teleology.

in fact essential. But even those biocentrists who have not committed themselves to an etiological account of teleological welfare should so commit themselves. The reason for this has mostly already been developed: in light of the subjectivist challenge, the biocentrist is burdened with doing more than appealing to the intuitiveness of the welfare of nonsentient organisms. Insofar as nonsentient organisms are teleologically organized in such a way that their teleology is nonarbitrary and nonderivative, the challenge can be met. The etiological account of teleology provides the resources to ground teleology in a way that is nonarbitrary and nonderivative. The fact that an etiological account of teleological welfare grounds or helps explain the plausible judgment that nonsentient organisms have a welfare provides a pro tanto reason to accept the account.

A-teleological Welfare?

Of course, it is open to the biocentrist to develop (a) an alternative account of welfare that is a-teleological or (b) a nonetiological account of teleology. I have not ruled out that possibility decisively, but any account of welfare must meet two requirements. First, it must meet the subjectivist challenge, providing an account of welfare that is subject-relative, nonarbitrary, and nonderivative. Second, it must be exclusive in just the right way, ruling out artifacts, ecosystems, and other entities that the biocentrist takes not to be bearers of moral status.

Ultimately, I do not think the etiological account of teleological welfare is exclusive in the way that biocentrists require. However, as the above arguments show, the etiological account of teleological welfare does meet the subjectivist challenge in ways that I think alternatives will not. Consider some examples of a-teleological accounts of the welfare of nonsentient organisms:

Integrity
X is good for organism O if and only if X promotes or maintains O's biological integrity.
Stability
X is good for O if and only if X promotes or maintains O's biological stability.
Growth
X is good for O if and only if X promotes or maintains O's normal growth.
Reproduction
X is good for O if and only if X promotes or maintains O's ability to reproduce.

For each of these accounts, firmly divorce in your mind any connection between what is purported to be good for an organism and the ends of the organism (and ignore for now the fact that at least some of these would allow for the good of ecosystems). Now, for each consider that they face an open-question-style objection: "Yes, X promotes the biological stability of O, but is it good for O?"[80] This seems to be me to be a perfectly good question, showing not only that "promoting biological stability" is not synonymous with "good for" but also not constitutive of it. This same kind of consideration doesn't seem to apply in the case of teleological welfare. To my mind, this is not an open question: "X promotes the ends of nonsentient organism O, but is it (pro tanto) good for O?"

Another way to put the point is that for any of these four accounts (or some combination of them), there is a question about why the purported constituent of welfare is in fact a constituent of welfare. Why should we think that normal growth is part of the good of nonsentient organisms? This open explanatory question is a way of pointing out the burden of having to provide some nonarbitrary account of welfare. I think this burden is extremely difficult to meet and probably explains why the biocentrists discussed above tend to lean on teleology in accounting for the welfare of nonsentient organisms. In any case, these issues are revisited in chapter 5, where I consider ways that these properties might be conjoined with teleology in an effort to show that artifacts can be excluded from being morally considerable.

Nonetiological Teleology?

There are nonetiological accounts of natural teleology. The best-developed of these accounts, often referred to as "autopoietic," "internal" or "organizational" accounts, ground the teleology of organisms in their being self-maintaining or self-organizing; the ends of a thing are grounded in or constituted by whatever it is self-organized to achieve (Mossio and Bich 2014). Claims about self-directedness, internal organization, or autopoiesis grounding or constituting teleological organization, and thus welfare, appear in the accounts of some biocentrists and others who think nonsentient organisms have a good of their own (Rolston 2003; Holm 2017).

Whether the autopoietic account will serve the purposes of biocentrists depends on whether it turns out that all living organisms will be teleologically organized on such an account and whether it is only living organisms

---

80. See Moore 2004.

which are so organized. Autopoietic accounts have an easy time with instant organisms. Since instant organisms are intrinsic duplicates of an organism, they will be self-maintaining in the same way that their counterparts are; the instant organism will have the same teleological organization as its noninstant counterpart.

Even if all living organisms are internally organized so as to be teleologically organized according to an autopoietic account of teleology, such accounts do not meet the requirement of being exclusive in the way that biocentrism requires. There are debates about whether, for example, the water cycle (Holm 2017; Toepfer 2012) and combustion processes (Holm 2017; Bickhard 2000) satisfy the requirement of autopoietic organization. Furthermore, as will be argued in chapter 5, many artifacts satisfy the requirements of being teleologically organized according to these accounts. This by itself isn't a bad result; artifacts *are* teleologically organized. However, one of the seeming virtues of the etiological account is that there are different sources of teleology for organisms and artifacts. Organisms are the result of *natural* as opposed to *artificial* selection, and this has been seen as key to explaining why it is that organisms but not artifacts are ultimately morally considerable. It is precisely this difference that Varner (1998, 69), for example, thinks does the work of excluding typical artifacts from being morally considerable.

Holm (2017, see also 2012), who defends the autopoietic account over an etiological one, thinks that biocentrists should accept this failure of exclusivity over the failure of etiological accounts to accommodate instant organisms. This is primarily because he thinks that it is extremely important to be able to accommodate instant organisms by including them among the domain of beings with a welfare. However, he also thinks that autopoietic accounts can avoid the conclusion that most artifacts are teleologically organized in the same way as organisms (Holm 2017, 1081–85). In chapter 5, I will return to the question autopoiesis and artifacts. I will argue that nothing is gained by adopting an autopoietic account of welfare or trying to use autopoiesis to exclude artifacts. Insofar as those arguments are sound, the challenge to biocentrism is sustained even if teleology is grounded in autopoiesis.

Autopoietic accounts also face a version of the open-question challenge raised against a-teleological accounts of the welfare of nonsentients. Yes, we may admit, O is a self-organizing system and X contributes to maintaining that organization, but is X good for it? Or, we may legitimately say, "Yes, O has parts and processes that self-maintain, but is that the purpose or end of those parts and processes?" The autopoietic account doesn't really explain why there is a connection between self-organization and teleology in the same way that an etiological account does by identifying ends with, literally, what a trait is selected for.

As I've been describing an etiological account of teleology, the sort of etiology relevant to teleology is a selection etiology. Lewens has defended an account of function on which *sorting*, as opposed to selection, is sufficient for grounding teleology (Lewens 2004, 126–27). A sorting process is one in which there is nothing akin to reproduction. As an example, he considers a screening process that is used in the development of pharmaceuticals: "Drug companies discover useful drugs simply by putting millions of randomly generated molecules through a series of chemical 'screens' where they are tested for some effect. The result is often the generation of a set of molecules well adapted to the tasks set by the screens, yet there is no selection process" (Lewens 2004, 127). Lewens goes on to argue that it is perfectly legitimate to talk about the purposes of the molecules that pass through the screens, and further, that ascriptions of teleology in this case are not best understood as being grounded in the intentions of the sorters but in the sorting process itself.[81]

Biocentrists might be uneasy about expanding the etiologies that generate teleology. As we will see in the next chapter, some assumptions biocentrists make about natural selection are essential to their strategy of exclusion; it is because wholes and organisms are not subject to natural selection that they lack a welfare grounded in teleological organization. Sorting is much easier to come by than selection. So if Lewens is correct, biocentrists who wish to, or are forced to, adopt a broader etiological account of teleological welfare that countenances interests grounded in sorting processes will have to rely on other resources besides some purported features of natural selection to carry out the strategy of exclusion.

Ultimately, the challenge to these nonetiological accounts of teleological welfare as a basis for biocentrism is that they face a dilemma: either they do not meet the subjectivist challenge, or they have the same consequences as etiological accounts that, I will argue, require that biocentrism be abandoned.

---

81. Lewens himself is not a biocentrist, and his focus is not on issues of welfare or moral considerability. Instead, his focus is simply on understanding the relationship between function ascriptions in artifactual and biological contexts.

ARTIFACTS, ECOSYSTEMS, AND THE DEATH OF THE ETHIC OF LIFE

4 THE STRATEGY OF EXCLUSION AND THE PROBLEM OF COLLECTIVES

Part I carves out space for welfare-based biocentric individualism by developing an account of the welfare of nonsentient organisms. In my view, an etiological account of teleological welfare, i.e., a teleological account of welfare where teleology is grounded in an etiological account of teleology, is not only the most plausible way to make sense of the welfare of nonsentient organisms (and the biological component of the welfare of sentient organisms), but also fairly represents and articulates the foundations for how many prominent biocentric individualists tend to understand or represent their view. The task of Part II is to articulate why biocentric individualism is untenable, why it must be rejected *despite* there being a plausible account of the welfare of nonsentient organisms.[82] Ultimately, sentientists are correct that nonsentient organisms are not morally considerable. However, this is not because such entities lack a welfare, but because that welfare is morally insignificant: it does not matter from the moral point of view.

The problem for biocentrism is a problem of exclusion. According to biocentrism neither biological collectives, such as biotic communities, ecosystems, or conspecific groups, nor artifacts, such as can openers, corkscrews, and (nonconscious) computers, are morally considerable. These are plausible exclusions from the domain of moral considerability, but this exclusion must come from somewhere. I will argue that there is no basis for this exclusion, that the biocentrist does not have the resources to exclude either collectives or artifacts from the domain of the morally considerable.

---

82. The arguments in this chapter and the material on the levels of selection were developed in some form in my dissertation (Basl 2011) and have since been further refined and published as an article (Basl 2017). I draw heavily on these, especially the latter, in what follows.

Biocentrists, to my knowledge, without exception have attempted to make sense of the relevant exclusion employing what I identified in chapter 1 as the strategy of exclusion. They argue that those things that are excluded by biocentrism are excluded because they lack a welfare; collectives and artifacts are not morally considerable because there is nothing that can make them better or worse off in the way that nonsentient organisms can be made better or worse off. Put another way, there is no way to make use of the strategy of extension to justify extension of moral considerability to nonsentient collectives or artifacts because there are no shared interests between members of the anchor class, nonsentient organisms, and members of the extension class, nonsentient collectives or artifacts.

Furthermore, it is the resources of an etiological account of teleological welfare that are often harnessed to help draw these boundaries. By grounding the sort of teleology relevant to well-being in natural selection and claiming that neither artifacts nor collectives are subject to natural selection, biocentrists have thought to have solved the problem of, or provided a basis for, exclusion. However, the etiological account of teleological welfare does not generate the exclusion class that biocentrists are committed to. In this chapter and the next, I argue that the problem cannot be solved by denying that collectives (this chapter) or artifacts (chapter 6) lack a welfare. If the etiological account of teleological welfare is correct, then both collectives of a certain kind and many artifacts have a welfare of the same kind and grounded in the same way as nonsentient organisms.

There is an alternative approach for biocentrists. They may accept that artifacts and collectives, like nonsentient organisms, have a welfare but argue that only the welfare of nonsentient organisms is normatively significant. This is different from the strategy of exclusion identified as part of the welfare approach, but it is theoretically available to the biocentrist. No biocentrist has, I think for good reason, attempted such a response to this problem of exclusion. In chapter 6, I briefly take up this possibility while arguing that, in light of the inability to solve the problem of exclusion, biocentrism must be abandoned.

Biocentrism, Teleological Individualism, and the Levels of Selection

Biological collectives, whether they be conspecific groups of organisms, biotic communities composed of organisms of different species, entire species, or ecosystems composed of organisms and abiotic components, are often described as having a health, a welfare, or at least, as being such that they can be made better or worse off. And holists have attempted to develop and defend accounts of ecosystem health (McShane 2004). Often, following Leopold, the welfare or health of ecosystems is understood in terms of their integrity or stability, and there are formal measures of, for example, ecosystemic stability.[83]

However, from the perspective of defending the moral considerability of ecosystems, it isn't enough to be able to identify or develop objective measures of something like ecosystem health, for two reasons. First, if the arguments from chapter 3 are correct, meeting the subjectivist challenge requires teleological organization. So the proponent of the welfare of biological collectives must argue that biological collectives are teleologically organized; stability or integrity, however measured, must be the end or goal of the relevant collection. Second, if those who identify some notion of ecosystem health are to employ the strategy of extension to ground the moral considerability of wholes, it must be the same kind of welfare that nonsentient organisms have, at least if they are to convince biocentrists to accept holism.

Biocentrists and other critics of holism have denied that biological collectives are teleologically organized, that the "interests" of collectives are the same as or of the same kind as those of nonsentient organisms (Goodpaster 1978; Sober 1986; Taylor 1989; Varner 1998; Cahen 2002; Sandler 2007; Odenbaugh 2010). The seeming goal-directedness of ecosystems, they have argued, is illusory, supervening on the goal-directedness of the individual organisms that make up the collective. Here is Cahen's overall assessment of the goal-directedness of wholes:

> Some ecosystems do appear to have goals—stability for example. There is a complication, however. Mere behavioral byproducts, which are outcomes of no moral significance, can look deceptively like goals. Moreover, on what I take to be our best current ecological and evolutionary understanding, the goal-directed appearance of ecosystems is in fact deceptive. Stability and other ecosystem properties are byproducts, not goals. (Cahen 2002, 123)

These critics are "teleological individualists." Teleological individualism is the view that individual organisms, even nonsentient organisms, are end- or goal-oriented systems, while biological collectives, such as ecosystems or

---

83. For a discussion of various accounts of stability in ecology and criticisms see Justus 2008.

conspecific groups, are mere assemblages of organisms. A maple tree grows upward and outward *in order to* soak up the sun, but not in order to provide resources for the other organisms in its environment. It is a commitment to teleological individualism that is the foundation for biocentrists' individualism, their identifying only individual organisms as bearers of moral considerability. Biological collectives are excluded from the domain of moral considerability precisely because they are excluded from the domain of the teleologically organized.

Teleological individualism has become something of a dogma in environmental ethics. With few exceptions, holists have not sought to undermine biocentrism by arguing that ecosystems have a welfare grounded in their teleological organization but have instead sought to ground the moral status of wholes in other ways that do not depend on welfare, e.g., by appeals to intrinsic value.

One reason, I think the primary reason, for this ceding of ground is a widespread view about the *levels* or *units of selection*. The question of the levels or units of selection, of which levels of biological organization are subject to natural selection, dates back to at least Darwin (see Darwin 1964; Sober 2010) and has attracted much attention since.[84] We have already seen one way that the issue of the levels of selection is important to the project of defending biocentrism. Recall the importance of the units of selection for etiological accounts of teleological welfare. The proponent of an etiological account of teleological welfare, unlike the proponent of an etiological account of function, must be able to identify organisms, as opposed to their traits, as the ultimate unit of teleological organization. There is something about an organism that makes it itself a unified, end-oriented whole; it is the bearer of traits, not merely a collection of them. Just as a clock differs from an unorganized pile of otherwise identical clock parts, an organism is unified in a way that a pile of traits is not. To accept that being a collection of teleologically organized parts is sufficient for being a teleologically organized whole would introduce a serious problem for the teleological individualist: Why should we not see the endocrine system as a teleologically unified whole among others occupying a similar space as other unified wholes? Why shouldn't we recognize an ecosystem as a teleologically unified whole? Appealing to a view about the level of selection, at least at first glance, seems to resolve these issues. It is true that an organism is a teleologically organized unit in way that the heart is not; hearts were selected

---

84. For contemporary discussions of the issue see, for example, Sober and Wilson 1998; Okasha 2006; Lloyd 2007.

for because of the contribution such organs made to the fitness of *organisms*. Hearts do not pump blood because they increase(d) the fitness of other hearts, but the fitness of the heart-bearer. The same can be said of many of the traits of nonsentient organisms: they exist as they do because of the contributions they made to the survival and reproduction of individuals in an ancestral lineage. So in answer to the question, "Why should we see an organism as something that is a teleologically unified whole as opposed to a mere collection of teleologically organized parts?" we can reply that it is because "organisms are the unit of selection, not their parts! Their parts evolve because of the contribution they make to the fitness of the organism as a whole!"[85]

The issue of the levels of selection is also central to the commitment to teleological individualism. Proponents of teleological individualism often rely on the view that it is *only* individual organisms, rather than groups or other biological collectives, that are units of selection, and it is for this reason that nonindividuals lack teleological organization.[86] Sober defends teleological individualism via a view about the levels of selection as follows:

> Darwinism has not banished the idea that parts of the natural world are goal-directed systems, but has furnished this idea with a natural mechanism. We properly conceive of organisms (or genes, sometimes) as being in the business of maximizing their chances of survival and reproduction. We describe characteristics as adaptations—as devices that exist for the furtherance of these ends. Natural selection makes this perspective intelligible. But Darwinism is a profoundly individualistic doctrine. Darwinism rejects the idea that species, communities, and ecosystems have adaptations that exist for their own benefit. (Sober 1986, 185)

In this passage, we see a commitment not only to an etiological account of teleology but also to a basis for teleological individualism grounded in a view about the units of selection.

---

85. While Agar doesn't discuss in much detail the debate over the levels of selection, he does argue that it is organisms as units of selection that help subordinate the teleology of organs and processes to the ends of the organisms as a whole (2001, 97–99), and he attempts to defend teleological individualism from the objection that genes might be the main unit of teleological organization (107–20).

86. It is in the defense of or commitment to teleological individualism that the we can see clearly a connection between welfare and natural selection; natural selection is the tool by which biocentrists attempt to make good on the strategy of exclusion.

Despite widespread acceptance, teleological individualism is false. Teleological individualists are committed to a view about the levels or units of selection that is, in Sober's words, "profoundly individualistic," but they are also committed to grounding teleology nonarbitrarily and nonderivatively. Careful attention to the question of the levels of selection reveals that one cannot jointly satisfy these three commitments. The ultimate upshot is that biocentrists cannot employ the strategy of exclusion; the commitment to individual organisms as the sole bearers of welfare cannot be justified.

A Trilemma for Teleological Individualism

There are three broad families of views about the units or levels of selection, and these views can be distinguished according to how they answer two questions:

1. The realism question: Is there a fact of the matter about the level at which selection operates or about which things are units of selection?
2. The monism question: Is selection confined to operating solely at one level or upon one unit?

A *conventionalist* about the levels of selection answers both the realism question and the monism question in the negative; a reductionist about the levels of selection answers both questions in the affirmative; those who endorse what is sometimes called "multilevel selection theory" and what I will call "multilevel realism" answer the realist question in the affirmative but the monism question in the negative.

Reductionist views have been the most commonly appealed to by teleological individualists for mostly obvious reasons, but in what follows I take up each view in turn, explaining why none can be relied on to ground teleological individualism. Afterward I will also make the case that the most plausible view— and the one that anyone committed to the welfare of nonsentient organisms ought to accept—is multilevel realism and articulate the implications of this for biocentrism.

Conventionalism

According to conventionalism, there is no objective fact about the level of selection (Kitcher and Sterelny 1988; Kerr and Godfrey-Smith 2002; Waters 2005,

200).[87] Instead, whether we describe a process as group selection, individual selection, genic selection, etc., depends on context. Assume that a population of organisms divides itself into pockets of individuals and that these individuals differ only with respect to whether they exhibit altruistic or selfish behavior.[88] Some groups, say by random assortment, contain only altruists, others only selfish individuals, others a mixture. Assume that altruistic individuals in pure groups have a fitness of three (they produce three viable offspring), while altruistic individuals in mixed groups have a fitness of two. Selfish individuals in pure groups have a fitness of one, but in mixed groups have a fitness value of three. For simplicity, assume that all groups are two-membered groups. We can assign a group fitness to each group by simply summing the total number of offspring in each group. The fitness of a purely selfish group is two, a purely altruistic group is six, and a mixed group is five. In such a population, altruistic groups are the fittest—they make the greatest contribution in terms of offspring (or expected number of offspring) to the next generation.

However, according to conventionalism, there is nothing special about this group-level perspective; even when there are populations that can plausibly be understood as divided into groups, there are mechanisms or tools by which we can translate group fitness into individual or genic fitness.[89] For example, whereas the fitness of an organism is defined in terms of its individual contribution to future generations, typically measured in terms of viable offspring, there is another sort of fitness, called "inclusive fitness," that is measured in

---

87. I'm glossing over important differences between various versions of conventionalism. Kitcher and Sterelny (1988), for example, occupy a strange middle ground between conventionalism and reductionism; they argue that the level of selection relevant to the evolution of a trait is a matter of convention, but give special place to the gene because an explanation in terms of genes can always be given, whereas, for example in populations without groups, a group-level explanation is not always available. Dawkins, while originally adopting a form of reductionism which was realist in nature (1989), later (1999) espoused a form of conventionalism similar to the version endorsed by Kitcher and Sterelny. (Thanks to an anonymous referee for pointing me to this feature of the later work of Dawkins.) For a discussion and challenge to views of this form, see, for example, Lloyd 2007.

88. The sort of altruism under discussion is *biological* as opposed to *psychological* altruism. Biological altruism is defined in terms of the fitness consequences of traits. A trait is, roughly, altruistic just in case it provides a benefit to another organism at a cost to the bearer of the trait. Notice that this doesn't require that organisms have minds (Sober and Wilson 1998).

89. These tools were first developed largely in response to the problem of altruism, the problem of explaining how traits that diminish the fitness of the individual that has them for the benefit of others could evolve. The existence of biological altruism seems to call out for explanations in terms of group selection. Kin selection theory (Smith 1964), inclusive fitness theory (Hamilton 1964a, 1964b), and game-theoretic approaches (Axelrod and Hamilton

terms of the average fitness of like individuals across a metapopulation. In the example above, where individuals differ only with respect to a single trait, we can find the inclusive fitness of altruistic or selfish individuals by looking at their average fitness across both mixed and pure groups. So the inclusive fitness of any altruistic individual is two and a half, while the inclusive fitness of any selfish individual is two.

Let us assume that the "altruistic" sharing behavior evolves in this population. Instead of explaining the evolution of altruism in terms of the fitness of altruistic groups, we can explain the evolution of altruism in terms of the higher inclusive fitness of individuals that are altruistic. This strategy is always available to explain the evolution of such traits by natural selection, so we can always give an explanation in terms of individual inclusive fitness for a given adaptation. The conventionalist recognizes that we might have different explanatory aims and employ different sorts of evolutionary explanations in light of those different aims. If it is easier to assign fitness to "groups" within a metapopulation and explain the evolution of a trait in those terms, given our aims, then so be it, but there is nothing privileged about those explanations.

What is the relationship between conventionalism and teleological individualism? On the one hand, the tools that conventionalists have employed to translate from one level of selection to another purportedly make it possible to view all natural selection as occurring at the level of the individual, as in the toy example. That sounds great for teleological individualism except that such a view immediately violates nonarbitrariness and/or nonderivativeness. If conventionalism is true, it yields teleological individualism only if we restrict the explanatory aims or interests relevant to deciding which naturally occurring entities are teleologically organized such that biological collectives are not units of selection. What could justify this restriction of explanatory interests? The answer will not be some fact about organisms, but some fact about us and our explanatory interests. Whatever those interests are, the welfare of nonsentient organisms is no longer objectively specifiable. We can't make sense of how nonsentient organisms are teleologically organized (and so what their welfare is) absent reference to our aims or purposes. Furthermore,

---

1981) were developed or employed as ways to understand selection processes that seem to be at the group level in terms of selection at a lower level. Instead of postulating group selection, altruism can be seen as evolving because altruistic individuals have higher inclusive fitness. For criticisms of the attempt to undermine group selection by appeal to these tools see Sober and Wilson 1998; Sarkar 2007; Sober 2010.

there is a sense in which their teleology, and so their welfare, *reduces to* our ends or aims. Their welfare becomes derivative.[90]

This is not to say that the choices conventionalists make about how to model selection are arbitrary in the sense of random. There may be basis in biological facts for explaining the evolution of a trait in a given context in terms of one level of another (for example, the population of organisms isn't in any way organized into groups), but conventionalism is by its very nature at odds with the sort of metaphysical realism that is inherent to teleological individualism. The teleological individualist thinks there really is something special about *individual organisms*; they and only they *really* are the units of teleological organization. The conventionalist thinks there is no such reality about the level at which selection operates. So individualists must simply insist on their favored perspective.[91]

Multilevel Realism

Multilevel realism is the view that there is a fact of the matter about the level at which selection operates and that it might, at least theoretically or conceptually, operate at all levels of organization from the gene up to the ecosystem. There are various forms of multilevel realism.[92] The most prominent

---

90. This happens to be a case where there is a failure to satisfy both nonarbitrariness and nonderivativeness for the same reason, but these conditions can come apart. For example, consider that we might attribute to the host of a parasite the end of nourishing the parasite. This end might be objectively specifiable in terms of the teleology of the parasite, but it will not satisfy nonderivativeness; this end really reduces to the parasite's ends. Similarly, a child, as children are wont to do, might attribute ends to an inanimate object. Those might be taken to be ends of the object itself, i.e., they are nonderivative, but they are arbitrary.

91. In "Ecosystem Health," Katie McShane (2004) raises doubts about whether our choices about what constitutes an ecosystem undermine claims about ecosystems having a health. She argues that just because our choices determine which things make up an ecosystem, this doesn't undermine the claim that whatever ends up being an ecosystem relative to our choices might have a health. Perhaps the biocentrist can similarly embrace conventionalism to avoid the criticisms just raised. But even if McShane is right that conventionalism doesn't undermine attributions of health or welfare, this is of no help to the individualists. This is because as a matter of convention, in at least some contexts, there is no problem describing selection as operating at the level of nonindividuals such as groups. Even if conventionalism doesn't undermine genuine teleological organization, it is too permissive to ground individualism.

92. Multilevel realism should not be confused with defenses of group selection that came before it, often referred to as naive group selectionism.

is Sober and Wilson's trait-group framework (1998).[93] The trait-group framework consists primarily of a criterion of grouphood and a set of definitions for selection at a given level. According to the criterion of grouphood, groups are *trait-relative*; two organisms might constitute an evolutionary group relative to one trait but not another. Organisms constitute a group relative to a trait when the fitness of those organisms (relative to that trait) depends on which variant of the trait other individuals have. Groups do not have to be groups of conspecifics. Organisms of different species can constitute a trait-group relative to some trait.

We can use the example developed in the context of discussing conventionalism to understand group-relative traits. In that example, the fitness of any individual depends on who it is grouped with. An altruist has a higher fitness when paired with another altruist than when paired with a selfish individual. Therefore, according to the concept of grouphood used in the trait-group framework, pairs in this population are trait-groups relative to the sharing behavior/trait.

The criterion for grouphood tells us when a population contains groups but not whether there is group selection for any given trait. That depends on the definitions of selection at a level. According to the trait-group framework, group selection occurs when there is selection among groups, i.e., when there is a difference in fitness between trait-groups. Individual selection occurs when there is selection within groups, i.e., when there are differences in fitness between individuals within groups (Sober and Wilson 1998, chap. 3).[94] Returning to our example, altruistic groups are fitter than pure groups, but within mixed groups—the only groups within which there is a difference in fitness—there is selection for individuals that are selfish.[95]

Multilevel realism is, by its very nature, inconsistent with the teleological individualist's commitment to organisms' being the sole unit of selection.

---

93. I've chosen the trait-group framework for the purposes of illustration, but the conclusions drawn generalize to any form of multilevel realism. For a comprehensive discussion of the problem of the units of selection see Okasha 2006. See also R. Wilson 2004.

94. If a population does not contain any groups, individual selection need not occur within groups; instead, it is defined in terms of difference in fitness between individuals in the population.

95. The example I've used to explain the trait-group framework appeals to individuals vs. groups, but the definitions generalize. "Individuals" and "groups" can be understood to represent particles and collections at any level of organization. For example, the individuals might be genes and the groups might be organisms, or the individuals might be organisms and the groups ecosystems. The trait-group framework is a framework for understanding selection at any level of biological organization (Okasha 2006; Sober and Wilson 1998, 96).

If multilevel realism is true, whether collectives such as groups, biotic communities, or ecosystems are teleologically organized (and thereby qualify as having a welfare) is ultimately an empirical question. Once we settle on how to understand what constitutes "a group," "a population," "selection at a level," etc., it will be up to biologists, ecologists, and the like to determine how often the relevant conditions for selection at a nonindividual level are or have been met. Some discussion of these issues will be taken up when discussing the implications of multilevel realism for biocentrism, but here what is important is that teleological individualists cannot adopt this view about the levels of selection. They must instead argue that the conceptual frameworks, such as the trait-group framework, endorsed by multilevel realists are somehow mistaken or inadequate for understanding natural selection and that instead we must adopt a form of reductionism.

Reductionism

Among teleological individualists, as well as among biologists and the public, reductionism, the view that only individual organisms are units of selection, is the most common view about the levels of selection. This is largely thanks to the popularity of Dawkins's *The Selfish Gene* (1989) and, within biology, to the lasting influence of George C. Williams's *Adaptation and Natural Selection* (1996) as well as the development of kin selection theory and inclusive fitness theory.[96] It also seems the most promising basis for teleological individualism simply in virtue of the way that it answers the realism question and the monism question.

There are many forms that reductionism takes depending on the motivations for the view. Historically, one of the primary motivations for reductionism comes from taking what Dawkins calls the "gene's-eye view" of selection.[97] Dawkins's picture of selection is one on which replicators compete with one another to pass on copies of themselves. Replicators are entities which pass on structural information from generation to generation. The copying

---

96. Within environmental ethics, Taylor explicitly cites Williams's work as showing that only individuals are units of selection (Taylor 1989, 7 n. 1). For a discussion of the role kin selection and inclusive fitness theory have played in debates about the levels of selection, see Sober 2010, chap. 2.

97. It is worth noting that this view has been widely criticized and is no longer widely accepted among biologists and philosophers of biology. Still, given the prominence of this view outside of biology, it is worth recognizing the challenges for adopting an etiologically based teleological individualism if one accepts such a view.

mechanism need not be perfect, but copies should be structurally similar and should themselves be capable of creating accurate copies through time. While other structures could potentially serve as replicators, the replicators of evolutionary biology are genes. Cells and bodies are machines, vehicles, or *interactors* that genes use to win out in competition with other genes, where "winning out" means producing more copies.[98] On this picture of selection, there is no room for individuals, let alone groups, to be seen as units of selection; they are merely tools. True, it is the interactors that are exposed to the elements, but behind the scenes the gene is at work. Phenotypes do well or poorly only insofar as they result in more copies of the replicators that gave rise to them. Therefore, the replicator enjoys a privileged place on this conception of natural selection.

Dawkins prefers this gene's-eye view of selection because he sees it as helping to solve the puzzle of altruism. There are cases where organisms seem to behave in ways that are costly in terms of reproduction but that benefit others. How could selection, which favors reproductive success, result in such altruism? Dawkins's answer to this was to appeal to what is good for genes of the same type. The gene doesn't care if its interactor does poorly so long as more copies of the replicator are passed on. Sometimes an interactor's behaving in a way that lowers its reproductive success will increase the copies of that gene in the next generation. Consider, for example, long-term parental care. From the gene's perspective, offspring carry copies of the replicator, and so the parent's sacrifice isn't costly to the replicator so long as it increases the number of replicators in future generations.

This view of selection can be contrasted with what might be called the "standard view." According to the standard view, selection should be understood in terms of the "Lewontin conditions."[99] Lewontin (1970), in a paper on the levels of selection, described the Darwinian principle of evolution by natural selection in terms of populations of entities that *vary in phenotype*, where phenotypes *vary in fitness*, and where phenotypes are *heritable*. According to Lewontin, these conditions are necessary and sufficient for evolutionary change by natural selection.[100] Unlike the gene's-eye view, the standard view

---

98. See also Hull 1980; Dennett 1995.

99. This expression is borrowed from Godfrey-Smith (2009), who contrasts the gene's-eye view and what I'm calling the standard view and argues in favor of the latter.

100. The Lewontin conditions are not, in fact, sufficient for evolution by natural selection (Brandon 1995). Even when a population has members that satisfy the conditions, if mutation rates are too high or there is sufficient evolution due to drift, the effects of selection can be undermined. Peter Godfrey-Smith has, helpfully, defended the Lewontin conditions as

is *level-neutral*; it includes no commitment to the level at which selection operates. Lewontin's paper includes a discussion of the extensive range of levels at which he thinks selection might operate: from genes, to cells, to organs, to organisms, to groups, and beyond.

There are some good reasons to prefer the standard view to the gene's-eye view. The main objection to the latter is that it is not fully general (Sober 2000; Gould 2002; Okasha 2006; Godfrey-Smith 2009). Darwin's conception of selection is neutral with respect to units of inheritance. Whatever mechanisms or underlying causes resulted in phenotypic variation, selection could operate. The phenotype of offspring might be the result of the blending of characteristics of parents and still there may be evolution by natural selection (Gould 2002). This is not so on the replicator view. As Godfrey-Smith explains:

> I introduced the case of a population in which there is variation in height, a reproductive advantage associated with tallness, and moderate but imperfect heritability of height. Initially, suppose that reproduction is asexual, so each individual has only one parent. Taller than average individuals produce taller than average individuals, but the variation in height is fairly evenly spaced. No two individuals are the same height, either within or across generations. So organisms in this situation do not "replicate" themselves; they do not "pass on" their structure or type. But evolution can certainly occur. (Godfrey-Smith 2009, 33)

Another way to understand the issue of generality is to consider the evolution of the genetic system itself. If the replicator view is the correct way to understand selection, genes must have come about by some other process (Godfrey-Smith 2000, 411).[101] But, presumably, replicators themselves evolved by natural selection. According to the standard view, the evolution of things like replicators can be the result of an evolutionary process of selection, where things that were replicator-like came about through random variation, and then selection favored this phenotype. The generality of the standard view speaks strongly in its favor. The world could have been very different; it could have lacked genes, and yet natural selection might have been an important evolutionary force, one resulting in a diverse tree of life.

---

a general kind of recipe for natural selection even though the addition of other factors can undermine selection (Godfrey-Smith 2009, chap. 2).

101. For an overview of challenges to the replicator view, see Godfrey-Smith 2000.

Whether one adopts the gene's-eye view or the standard view, one cannot successfully use either view as a basis of teleological individualism. Adopting the gene's-eye view seems to immediately preclude a commitment to individualism in the sense we have been discussing. It is genes that are the primary unit of selection. So it will be some component of genes that is teleologically organized toward achieving the ends of the gene.[102] But the teleological individualist is concerned with carving out space for individual organisms. The biocentrist does not think that it is genes, but the organisms that have genes, that are morally considerable. On first glance, reductionism is actually at odds with teleological individualism.

Are organisms really precluded from being teleologically organized if we adopt the gene's-eye view? After all, interactors have instrumental ends or purposes. Their ends are to interact with the world for the purposes of propagating copies of genes. Why can't teleological individualists adopt the gene's-eye view and recognize organisms as teleologically organized interactors? They might be forced to accept that genes are teleologically organized units, but their teleological organization substantially overlaps with the teleological organization of their interactors, or so it seems. So, for most intents and purposes, it is fine to accept that genes are teleologically organized units along with individual organisms. Teleological individualism is perhaps diminished, but not badly so.[103]

The problem is that nothing precludes collectives such as conspecific groups or multispecies communities from serving as interactors. The gene's-eye view was developed, in part, to resolve the problem that sometimes groups seem to be units of selection; on the gene's-eye view, groups can be interactors. Take our toy example; we can understand pairs as units or interactors, working together so that more copies of their genes are passed on to the next generation. The idea of group interactors is even more clear in real-world examples

---

102. There is an interesting question about how to understand genes as being teleologically organized. What exactly is it that is so organized? There is, as far as I know, little discussion of this since most teleological individualists with an interest in these issues typically talk as if it is individual organisms that are teleologically organized even if they adopt a form of reductionism. Lewontin (1970) discusses molecules as units of selection. Perhaps we should understand the molecular structure of DNA as that which is teleologically organized. Thanks to an anonymous reviewer for raising this question.

103. Agar argues that there is a way of construing selection on the gene's-eye view such that the genes are teleologically organized toward the ends of the organism (Agar 2001, 107–20). If he is correct, the proponent of the gene's-eye view can eliminate seeing the gene as having a welfare. However, Agar's approach is subject to the problem that there will be nonindividual interactors, as discussed in what follows.

like ant colonies or bee hives. Teleological individualists hoping to ground their view in the gene's-eye view must find some way to exclude these collective interactors.[104] Again, there doesn't seem to be a ready way to draw nonarbitrary lines to exclude nonindividual or collective interactors while including organisms. A choice must be made between teleological individualism and nonarbitrariness.

While the standard view doesn't immediately entail that organisms are the sole units of selection, neither does it rule out that possibility. However, teleological individualists who wish to defend their view while adopting the standard view must provide some reason for thinking that, as a matter of fact rather than as a conceptual consequence, it is individuals that are the sole unit of selection.

In *Adaptation and Natural Selection*, George C. Williams argues that we should understand selection as operating at the level of individual organisms rather than groups on grounds of parsimony. He claims that the concept of adaptation is "onerous," to be invoked only when necessary. Furthermore, he claims that higher-level selection is a more onerous concept than lower-level selection. So higher-level selection explanations should be invoked only when it is impossible to explain the existence of a trait in terms of lower-level processes. In his words:

> The ground rule—or perhaps *doctrine* would be a better term—is that adaptation is a special and onerous concept that should be used only where it is really necessary. When it must be recognized, it should be attributed to no higher a level of organization than is demanded by the evidence. In explaining adaptation, one should assume the adequacy of the simplest form of natural selection, that of alternative alleles in Mendelian populations, unless the evidence clearly shows that this theory does not suffice. (Williams 1996, 4–5)

After articulating this doctrine, Williams goes on to discuss adaptations, such as altruistic ones, that seem to call out for explanations in terms of higher-level processes. He tries to explain each away by offering an alternative explanation in terms of lower-level processes,[105] and, as noted above,

---

104. McShane (2014) has, independently, developed a similar criticism. Agar (1997, 70–71) also briefly considers this problem.

105. See, for example, Williams 1996, chap. 7.

there are various tools that one might use to explain selection in terms of individuals.[106]

For purposes of argument, let's just assume that for any trait we wish to explain in terms of natural selection, there is some way to explain it in terms of individual selection. Does parsimony give us any reason to prefer that explanation to the higher-level explanation?[107] It does not. To see why, first distinguish between two different contexts. In some contexts, even those who disagree with reductionism will agree that a trait probably evolved by selection at the level of the individual. This is because different conditions must be met for group selection to occur than for individual selection to occur. Group selection, for example, requires groups of organisms that vary with respect to some heritable trait. In such a context where the conditions necessary for group selection are not present, we should prefer a lower-level selection process to a higher (though of course the trait might have evolved by something other than natural selection, random genetic drift, for example). But, in this context, it isn't parsimony that gives us a reason to prefer one selection hypothesis to the other; it is simply a matter of group selection not being a viable hypothesis at all.

Consider instead another context, one where proponents of group selection think the conditions necessary for group selection are present. In this context, does parsimony give us reason to prefer the lower-level selection hypothesis? While there are a variety of views about how best to understand what parsimony is, it isn't obvious in such a case that one of these hypotheses is more parsimonious than the other. To say that something evolved by group selection is not to propose anything metaphysically burdensome; group selection isn't a different process than individual selection, it is the same process operating on groups simply because there are groups that meet the Lewontin conditions. If those conditions are met, then the groups will be subject to natural selection, and under the right conditions, that group selection will result in evolution.

Of course, the proponent of reductionism might argue that those conditions are never or hardly ever satisfied, or that group selection is relatively weak compared to individual selection, and so there is never or hardly ever evolution

---

106. The same tools available to the conventionalist, inclusive fitness, game theory, etc. are available to the reductionist to explain selection at some higher level in terms of individuals or members at some lower level of biological organization.

107. For a recent overview and discussion of parsimony reasoning in biology, see Sober 2015. Sober argues that parsimony considerations don't *generally* tell in favor of hypotheses. Instead, the heuristic value of parsimony is limited to specific contexts.

by group selection. But these seem to be empirical claims that might be advanced against a particular selection hypothesis in a given case rather than reason to accept reductionism on the basis of parsimony. More importantly, even if higher-level selection forces are relatively weak or if the conditions are not often met, this is antithetical to teleological individualism, especially given the purposes to which it is often employed. It turns out that reductionism of this form doesn't justify teleological individualism and so doesn't provide part of the justification for the view that only individual organisms have a welfare. For those teleological individualists who have thought that something about the nature of selection itself precluded the teleological organization of biological collectives, this version of reductionism is of no help.

This result is unsurprising given that the Lewontin conditions are level-neutral. If we accept something like these conditions, whether reductionism is true is a contingent, empirical matter. Teleological individualists can, of course, wishfully bet on the facts coming out in a way that supports their view, but it is worth noting that this would be a radical departure from how teleological individualists have tended to argue that biological collectives are not teleologically organized. More importantly, it seems that this empirical bet is a bad one, at least if the bet is that there are no instances of evolution by natural selection at levels above the individual organism. As will be discussed in the next section, proponents of multilevel realism claim to have at hand various instances of traits that have evolved by higher-level selection processes.

Implications of the Trilemma

What follows from the trilemma for teleological individualism? Biocentrism is, strictly speaking, false. Careful attention to the issue of the levels of selection shows that one of the etiological account of teleological welfare or teleological individualism must be rejected, but both are essential to biocentrism. The same problem undermines criticisms, such as those of Cahen and Sober, who seem happy to grant that nonsentient organisms are teleologically organized but deny that collectives are.[108] There is just not space for one to see natural selection as grounding teleology in a way that is appropriately independent of our aims and interests while at the same time seeing natural selection as something that operates only on individuals, not on collectives.

---

108. Sober, to my knowledge, no longer endorses the view that only individual organisms are teleologically organized in the relevant sense.

While biocentrism is strictly false, the trilemma doesn't tell us in which of three ways biocentrism is false. It could be false because nonsentient organisms lack a welfare, because it is not *only* nonsentient organisms but also collectives that are morally considerable, or because the welfare of nonsentient organisms is not morally significant. Which of these is the right response to the trilemma depends in part on which view about the levels of selection is ultimately correct. For example, if conventionalism is true and, given the arguments from previous chapters, the etiological account of teleological welfare is the best bet for grounding the welfare of nonsentient organisms in a way that meets the subjectivist challenge, the only option is to recognize that the subjectivist challenge cannot be met; attributions of welfare to nonsentient organisms are undermined. They must be explained as, for example, metaphor or must just be our interests in disguise. However, on views such as reductionism and multilevel realism the subjectivist challenge can be met, but the relevant bearers of teleological welfare will include nonindividuals or will not include individuals at all (because genes might be the only unit of selection).

In my view, the biocentrist should ultimately accept that some nonsentient collectives have a welfare in the same way that nonsentient organisms do. But the claim that such entities have a welfare presupposes that multilevel realism is the most plausible view about the units of selection, i.e., that there are good reasons to accept that the answer to the realism question is yes and the answer to the monism question is no.

With respect to the realism question, most proponents of conventionalism (and many reductionists) have found it significant that any selection that we care to identify as higher-level selection, such as group selection, can be modeled at a lower level. In other words, no matter at which level you find selection, there is a mathematically equivalent model at a lower level of selection that is predictively equivalent—one telling us that the trait that has evolved would be expected to evolve. Tools such as inclusive fitness help to generate these mathematical equivalences. But what are the true implications of this fact?

Sober and Wilson have convincingly argued that any inference on the basis of kin selection, inclusive fitness theory, and game theory to the view that there is no group selection commits the "averaging fallacy." Each of these strategies for addressing the problem of altruism appeals to an average fitness that is then assigned to individuals so as to make it seem as if we can understand a bit of natural selection in terms of higher individual fitness rather than in terms of group fitness. In the case of kin selection and inclusive fitness theory, it is the average contribution a gene makes to the gene pool in terms of copies of itself. On game-theoretic approaches, it is the average fitness of a trait across n-sized groups with that trait. Averaging in this way, Sober and Wilson explain, obscures the actual process and dynamics involved in the evolution of altruism; there are groups with some fitness, and sometimes groups of altruists do better than groups of selfish individuals even when selfish individuals are fitter within groups.[109]

More recently, Sober (2010; see postscript) has taken up what mathematical equivalence means with respect to conventionalism. Mathematical equivalence, he argues, allows one to give a model or explanation of a trait *in terms of* individuals or genes, but just because a model can be given in terms of, for example, individuals, does not mean that it is a *model of* individual selection. As Sober notes, the fact that groups supervene on individuals is enough to ensure that an explanation of the evolution of a trait can be given in terms of individuals if it can be given in terms of groups. What matters for the realist is whether the fact of grouphood makes a difference in the evolution of a trait. The fact that two models of the evolution of a trait are mathematically equivalent doesn't tell us that groups aren't relevant to evolution. In fact it can often obscure their importance.

When it comes to issues of teleological organization, the specific details of etiology matter quite a lot. Even in the toy example of the evolution of altruism, this is apparent. Why are some organisms altruistic? It is because groups that have more altruists do better than those with less; altruism exists because of the effect it has on the group. If we translate this to inclusive fitness talk, we miss something important. Yes, altruists have a higher inclusive fitness, but altruism doesn't help those that get stuck with selfish individuals; it is bad for the altruist to be an altruist when nobody else is an altruist. Translating to a mathematically equivalent model can obscure important causal interactions and relationships.

Perhaps the importance of the specific causal structure of events is more apparent in debates over how to understand actual traits that seem altruistic. Take the familiar example of the barbed stinger of the honeybee. The stinger is barbed in such a way that stinging, typically, results in the death of the honeybee. Why did the barb evolve? One seemingly plausible answer is that, in ancestral populations, having the barb increased the fitness of groups of bees that had group members with barbed stingers even while lowering the fitness

---

109. Sober and Wilson do not claim that there is no use for inclusive fitness or kin selection theories. Rather, they see these as useful tools, but they see models using inclusive fitness or kin selection as models of higher-level selection.

of the bee with the barbed stinger. In other words, the barb evolved, or exists now, because of its fitness effects for the group rather than the individual.[110] Perhaps that is the wrong causal history, perhaps barbed stingers, somehow, were good for the individual that had them or were genetically correlated with some other trait that enhanced individual fitness and so stuck around despite offering no advantage. These different causal histories seem relevant to the issue of teleological organization. If one causal story is correct, hives are tele-ologically organized and barbed stingers have group-relative ends; if the other is correct, they do not. The fact that we could describe the evolution of the barbed stinger, come what may, in terms of inclusive fitness obscures exactly what we care about when thinking about teleological organization: the details of causal history that explain why a trait exists as it does in the entity or entities of which it is a part.

It is for this same reason that we should reject a priori justifications for reductionism. The right conception of selection is level-neutral, and what matters is the actual causal history of a trait, whether it evolved because it was good for collectives or individuals. If it is only individuals that are units of selection, and thereby teleologically organized, this must be because, as a matter of causal history, there were no instances where a trait evolved because it provided a benefit to a collective. At the very least, proponents of an etiolog-ical account of teleological welfare should recognize that it is empirical details that will ultimately settle questions about the level of biological organization at which there is teleological organization.

Furthermore, once we recognize that the debate between multilevel realists and reductionists should be seen as largely empirical, the most plau-sible answer to the monism question is no. Contemporary defenders of group selection, while they disagree about exactly how to define groups and what group fitness dynamics must hold for group selection to occur, have hosts of examples where they think the relevant conditions for group selection have been met (see, for example, Wade 1976; Goodnight and Stevens 1997; Sober and Wilson 1998; R. Wilson 2004). Once we see that translating from a tradi-tional concept of fitness to some other concept doesn't undermine the relevant causal relations that constitute group selection, there are many cases where it seems uncontroversial that groupings are causally implicated in the selection of a trait. Species like the honeybee that have populations with sterile castes

---

110. There is a sense in which this example is misleading since the worker bees in a hive are a sterile caste; they don't reproduce. So they don't have a fitness at all. However, this makes it all the more clear that their contribution to hive evolution is via contribution to fitness of the group.

(subclasses that cannot reproduce) have traits whose existence cannot be (causally) explained because it was good for the particular individual ancestors that had it. There are other eusocial insects, such as various species of ants and wasps, that have similar castes (Crespi and Yanega 1995). While most of the literature on the evolution of altruism and multilevel selection concerns groups where "group" is understood as groups of conspecifics, there is recent work concerning what is referred to as "community selection" which involves organisms of different species (Swenson, Wilson, and Elias 2000; Swenson, Arendt, and Wilson 2000; Wilson and Swenson 2003).

Which Collectives

The difficult question is not whether collectives are subject to natural selection, but which ones are subject to natural selection and have evolved because of it. And while answering this question is not important to the truth of biocentrism—it is false however we answer these questions—it is important for seeing what exactly proponents of etiological accounts of teleological welfare are committed to and, ultimately, for seeing what some of the consequences are if we must recognize everything that has a teleological welfare as having moral status. I close this chapter by discussing which collectives seem most likely to be candidates for being teleologically organized and explain why these are not likely to be the sorts of things that holists see as having moral status. To make progress, I will consider three different conceptions of grouphood and group selection in order of permissiveness, or in how easy it is for a community of organisms to count as having evolved because of selection at the level of the community.

Collective Sorting

Some experimental work in biology attempts to demonstrate the possibility and efficacy of community selection by drawing a lesson from Darwin and employing what I will call "collective sorting."[111] Darwin appealed to the demonstrated capacity to change species in radical ways via artificial selection to help make the case that speciation is, in principle, possible via natural selection (Darwin 1964, chap. 1). Humans, by selectively choosing which individual organisms are allowed to produce offspring, can shape the traits

---

111. The authors simply refer to the process as community selection, but that obscures the distinction I'm trying to draw.

of future generations, and those generations can differ radically from distant ancestors. Nature, over a long period of time, can yield similar changes because there is variation in individual fitness. In Darwin's case, individual organisms are chosen from a population of individuals and bred because they have some desirable traits. That is, parent generations are chosen by "individual sorting." In some community selection experiments, a community phenotype is chosen and whole communities are selectively chosen from a larger population of communities on the basis of that phenotype. This is "collective sorting." Collective sorting concerns the level of discrimination when selecting parent generations. Selective breeding of, say, farm animals involves sorting by individuals; individual animals are chosen or selected to serve as a parent generation.[112] In collective sorting, rather than choosing or selecting individual organisms as a parent generation, a parent generation is populated by choosing a collective of various organisms to use as a parent generation.

There are a host of experiments in biology that employ collective sorting to demonstrate the possibility of community selection or, at least, an artificial analogue. In one paper, Swenson et al. attempted to demonstrate community selection in two types of communities: soil communities and aquatic communities. In the soil community experiments, containers of sterilized soil were inoculated with soil samples containing naturally occurring soil communities. Seeds were then planted in these soil samples to serve as the basis for the phenotype to be selected. These containers were divided into various control and experimental groups and subjected to different selection regimes. In control groups, soil samples were collected from random containers to populate a new generation of containers. In the experimental group, containers were chosen on the basis of above-ground biomass; containers that resulted in greater above-ground biomass were sampled to populate a new generation of containers. This was done repeatedly to create a lineage of containers using these sampling procedures.

In the aquatic community experiments, pond water containing a community of microscopic organisms is collected and used to populate an initial generation of test tubes. The pH of each tube is determined, and then the tubes are subjected to artificial selection, with samples being drawn to populate future generations. Again, the starting populations are divided into a

---

112. Collective sorting techniques have also been deployed in agricultural contexts for the purposes of improving animal welfare (see, for example, Muir 1996, 2003). For example, groups of hens that have higher production than other groups can be chosen as a parent generation for future groups of hens. The thought is that groups with higher production likely have better group dynamics, fighting less, for example (see Muir 2003).

control group and into various selection regimes. One set of communities was selected for high pH and another for low pH, with new tubes being founded using samples from tubes with the desired phenotype. This resulted in two different lines of communities subject to different selection regimes and one line of communities chosen at random.

Swenson et al. found a response to selection in all of the selected lines in these experiments. For example, later generations in the lines selected for above-ground biomass tended to produce more above-ground biomass than earlier generations (and the control group), and later generations of aquatic lines selected for low pH had lower pH than their ancestors (and the control group). In their discussion of the experiments, they claim that "The response to ecosystem-level selection in our experiments stands in contrast to theoretical models that have led many to reject higher-level selection as an important evolutionary force" (Swenson, Wilson, and Elias 2000, 9111). Not only do the experiments purport to demonstrate the possibility of community selection, but also to show that its influence might be greater than otherwise expected.[113]

In another experiment, Swenson, Arendt, and Wilson selected communities on the basis of their ability to degrade an environmental toxin, 3-chloranaline (Swenson, Arendt, and Wilson 2000). Test tubes were incubated with communities of microorganisms sampled from pond water as well as with the environmental toxin. Tubes were divided into eight lines and each tube was measured for the amount of toxin degraded. Four lines were subject to selection such that parents were selected according to which tubes degraded the most toxin; tubes that exhibited the most degradation of toxin were sampled to start new tubes. The other four lines, controls, had tubes selected at random to start new generations.

In three of the four selected lines, there was a systematic increase in how much toxin the communities degraded. On average, later generations within a line were better degraders than earlier generations. The fourth selected line showed no response to selection. Swenson et al. hypothesize that this could be due to sampling error resulting in an initial community that lacked the relevant community composition to then be subject to selection. In the control

---

113. In some places, including here, Swenson et al. use ecosystem selection and community selection interchangeably. However, in another paper Swenson et al. distinguish between the two, defining an ecosystem as "a system formed by the interaction of a community of organisms with their environment" (Swenson, Arendt, and Wilson 2000, 566). Ecosystems in this sense will not be units of selection for reasons discussed below.

lines, there was no systematic response to selection over the course of the experiment (30 generations of tubes).[114]

It is implicit in the conclusions drawn from these experiments that community sorting is sufficient for or identical to community selection. In each case, collective sorting occurs, a response to selection is observed, and it is claimed that community selection is possible. What seems to make individuals members of a community is simply that they are bound together enough to be subject to collective sorting—that whether they live or die simply depends on whether they are in some way associated enough to be collected up to form a next generation.

Understanding community selection in terms of collective sorting is a mistake. Let's imagine that I have a very large collection of very large aquaria and in each I keep one shark. Let's say that each shark has several remoras attached it to it that provide services for the shark and feed on scraps of the shark's meals. Let us also suppose that the population of remoras is genetically identical. One day I decide that I want to create the ultimate predator and begin by selectively breeding sharks for the sharpness of their teeth. I choose the males and females with the sharpest teeth and breed them. Of their offspring, I also choose the sharks with the sharpest teeth and breed them. I continue this experiment for many generations and am careful to avoid too much inbreeding. Lo and behold, I am successful: my latest generation of sharks has the sharpest teeth ever recorded.

What is it that I've been successful in doing? It seems to me that I've used individual selection in order to artificially influence the evolution of these sharks. But, as a matter of fact, every time I chose a shark to breed, I urged it, via a series of underwater tubes, into a breeding tank. In each case, the shark was accompanied by its remora. This is just to say that I've engaged in collective sorting to generate parents for the next generation of shark. If collective sorting always amounts to community selection, my experiments constitute community selection. But this seems entirely implausible.

The problem with understanding community selection in terms of collective sorting is that communities or groupings are too easy to come by; the notion of group is too easily satisfied and doesn't capture the relevant sorts of causal relations that are important for natural selection, at least when we care about the role that selection plays in grounding teleological organization. A more plausible understanding of community selection requires a more restrictive notion of grouphood.

---

114. Note that a generation of tubes may constitute many generations of microorganisms.

My silly example with the sharks is illustrative, but not necessary. If collective sorting is sufficient for community selection, then there is unlikely ever anything other than community selection. When a lion kills a slow gazelle, it also very well might be engaged in community sorting. The slow gazelle, as well as the survivors, may have parasites and certainly is covered in bacteria of different species. Yet this seems like a paradigm case of individual selection. Collective sorting may be a very useful tool and may help us to cultivate communities that have desirable traits without having to determine which individuals in those ecosystems are bringing about the desirable effects (Swenson, Arendt, and Wilson 2000), but collective sorting is not community selection; collective sorting that results in evolution doesn't entail that the collective is teleologically organized.

Correlated Interaction

A more restrictive and orthodox notion of groups or communities is to define them, as in the trait-group framework, in terms of *correlated interactions*.[115] To define groups in terms of correlated interactions is to define them in terms of the mutual sensitivity of fitness of one individual to the phenotype of another individual.[116] The definition of groups as trait-groups is a definition of groups in terms of correlated interaction. Recall from previous examples involving trait-groups that the fitness of selfish individuals depends on whether they are paired with another selfish individual or with an altruist, and vice versa.

Defining groups in terms of correlated interactions is more plausible than defining groups in terms of being subjects of collective sorting. At the very least, it avoids the counterexamples raised in the previous section. However, it also undermines the evidential import of the community selection experiments discussed. If we define groups in terms of correlated interactions, then the community experiments discussed above do not justify the conclusion that they are examples of community selection. To see why, consider a modified version of the 3-chloranaline experiment. For simplicity sake, let's assume that tubes contain only two species, A and B. Each tube contains some amount of A organisms and B organisms, and tubes that degrade the most toxin are selected as parents for future generations. In this example, there is collective selection. Is there community selection? This depends on what is going on

---

115. Glymour (2008) identifies correlated interactions as the orthodox view about how to define groups.

116. For a more precise definition see Glymour 2008.

inside the tubes. There are many ways that degradation may occur within the tubes. It may be that A individuals begin a chemical process that degrades 3-chloroanaline and B individuals complete the process. Perhaps if there were no B individuals in a tube, A individuals could not even start the process. Or perhaps the A's efficiency is, in some way or other, impacted by the number of B individuals in the tube. If this is what is happening, then there is reason perhaps to see the community as causally interacting in ways relevant to there being community selection. These organisms, for example, satisfy the definition of being a trait-group or trait-community.

However, the degradation may occur another way. It may be that all that matters to degradation is the number of A individuals in the tube. Let's assume that, given how quickly tubes are chosen to form new generations, the number of B individuals does not constrain the number of A individuals in a tube; reproductive rates of B individuals never influence how many A individuals are in the tube. In this case, there are no mutual fitness interactions relative to degradation. In other words, the results of the experiment do not justify the conclusion that there has been community selection because there is no evidence of correlated interaction.[117]

What are the implications of conjoining something like the trait-group framework with an etiological account of teleological welfare? As was noted, this depends on the empirical details, but we can learn something by noting some of the necessary conditions that would have to be met for a community to be teleologically organized. There is no group selection without groups. So a necessary condition is a metapopulation that is divided into groups defined by correlated interactions.[118] A community will meet this requirement only if there are fitness interactions that hold between all or most members of the community that do not hold between members of that community and organisms outside that community (Glymour 2008).

The ecosystems and species of traditional holism are very unlikely to ever satisfy the condition for being a trait-group or trait-community. Species are just too geographically separated for the relevant interactions to hold between all and only members of the species. The exception to this would be species that are extremely small and co-located, cases where, for example, a species

117. For some experiments that purport to demonstrate the possibility of community selection where groups are defined in terms of correlated interactions see Goodnight and Stevens 1997. For a criticism of what these experiments show, see Glymour 2017.

118. There is a further challenge to say exactly what constitutes a population or metapopulation. How are we to, for example, determine whether two organisms are part of the same population? For a discussion see Millstein 2009.

is endangered and all the remaining members of the species are in the same place. In such conditions, it is at least possible that all the members of the species constitute a trait-group.[119] Similar considerations apply to ecosystems. The ecosystems of traditional holism are large swaths of land with organisms of many different species. It is hard to imagine what trait we could identify such that how every member varied with respect to that trait affected the trait-relative fitness of every other organism that makes up the community. Furthermore, even if all the organisms of an ecosystem were connected via the relevant fitness interactions, it is hard to see what would constitute the alternative community that was less fit. For evolution by natural selection there must be variation in fitness, and an ecosystem can't vary in fitness with respect to itself. Of course, every ecosystem is made up of subsystems, and perhaps some of these might satisfy the conditions of being a trait-community, but there is no reason at all to expect the ecosystem to be a collection of trait-communities. There is certainly a possible world where every large ecosystem is composed of many trait-communities where there was, historically, selection for some trait-communities over others resulting in a community of communities that are teleologically organized. It's a beautiful world, but it is very unlikely that it is the actual world.

Understanding higher-level selection in terms of some version of multi-level realism that defines groups in terms of correlated interactions might seem appealing to the biocentrist given that, at least, it is far less permissive than an account that appeals to collective sorting. Unfortunately, accounts of group selection that define groups in terms of correlated interactions face serious problems. One problem concerns how those who employ a correlated interaction definition of groups typically understand or define group selection or selection at a level. Recall that according to the trait-group framework, a group is a trait-group, defined by correlated interactions, and selection at the group level occurs when there is differential fitness between trait-groups, whereas individual selection occurs when there is differential fitness between individuals within groups. The problem is that this definition of grouphood in combination with the proposed definitions of selection at a level yields implausible consequences. In particular, under certain conditions, the framework

---

119. There is a literature on "species selection," where species are viewed as units of selection, in some sense, and where species selection is seen as an important evolutionary force, one that helps to explain, for example, patterns of speciation (see Jablonski 1987, 2008; Lloyd 1994; Lloyd and Gould 1993). While species selection, in this sense, might be explanatorily useful, it doesn't seem that it would ground claims about the good of entire species. For discussion see Powell 2011.

entails there is no selection at a given level even when it seems clear that there is. A helpful way to understand the problem, which I have called "the disappearing selection problem," has since been developed by McLoone (2015) in terms of the Stag Hunt game.

The Stag Hunt game, like the other games of game theory, such as the Prisoner's Dilemma, is a tool for modeling the outcomes of interactions between players or participants in the interaction. Games are defined by the nature of interactions between the players, in terms of, for example, whether players interact only once or multiple times, which traits or behaviors the players might exhibit, whether players may communicate, etc., and by the payoffs for the different types of interactions between players. In Stag Hunt, we consider a group of players who will sort into pairs to hunt. Players in the game either opt to hunt stag, which cannot be hunted alone, or to hunt alone, in which case the player hunts hare. The strategy of hunting stag or hunting hare is chosen by (or assigned to) each player prior to being sorted into pairs, and there is no chance to change one's strategy. Payoffs are constructed such that the best possible outcome for an agent is to hunt stag when your partner also opts to hunt stag. However, the worst outcome is to opt to hunt stag when your partner opts to hunt hare. The value of hunting hare is stable no matter what your partner chooses. This payoff structure captures the idea that hunting stag, if successful, is a more efficient way to gather resources, but if one attempts it alone, one gets nothing. Hunting hare, on the other hand, is a steady way to gather resources. Often, games are described such that the players of the game are agents who make choices over various behavior options, but this need not be so. Game theory is a useful tool for modeling evolution by natural selection because the payoffs can be understood in terms of fitness; interactions between players need not be interactions by choice, but can be random pairings, and the "choices" need not be choices at all, but just different traits. So Stag Hunt can be described so as to remove all agency. Just imagine there is a population of nonsentient organisms, some of which are genetically predisposed to prey on something they can't kill on their own and some of which are predisposed to prey on something they can hunt alone. They are randomly sorted into pairs. Those that are predisposed to hunt difficult prey do very well if their partner is also so predisposed but do very poorly otherwise. Those that are predisposed to hunt easy prey do all right no matter who their partner is.

Consider a population of six organisms playing Stag Hunt with one pair of stag hunters, one pair of hare hunters, and one mixed pair. Given the payoffs as described, the stag hunters do better than both other pairs. All the hare hunters, whether in the pure group of hair hunters or the mixed group, do the same, but because the stag hunter in the mixed pair fails to hunt, the mixed pair does the worst.

The problem for understanding selection according to the trait-group framework is that there is no selection operating at all in this population. Even though individuals in Stag Hunt are sorted into pairs, these aren't necessarily groups in the sense defined in terms of correlated interactions. Notice that the fitness of stag hunters depends on their pairing, so relative to hunting behavior, stag hunters can form trait-groups, but the fitness of hare hunters is stable across pairings, so they never form trait-groups. So in this population, there is one trait-group, the pure group of stag hunters, and then a bunch of individuals. Group selection is defined as selection among groups, but there is only one group. So there is no selection among groups. Individual selection is defined as selection within groups, but the only group is pure, and so there is no difference in fitness between individuals within that group, and so no individual selection. Intuitively, there is selection for groups of pure stag hunters: the pure group of stag hunters will leave the most off-spring. However, there is a sense in which there is individual selection for hare hunters. All the hare hunters do better than the stag hunter stuck in the mixed group. If your intuitions about whether and where there is individual or group selection differ, that's fine, but what seems clear is that natural selection will operate on such a population in some way, yet according to the trait-group framework there is no selection at all.

The definition of groups in terms of correlated interactions could perhaps be maintained by pairing it with a different set of definitions for selection at a level, but the definition of groups in terms of correlated interaction faces other challenges. There are serious epistemic hurdles to identifying whether the relevant causal interactions necessary for grouphood can be reliably identified. Think of what one would have to know in order to say that this or that trait is there because of something it does for a group: One would have to be able to identify traits where variation in that trait contributed differentially to fitness depending on which variant of that trait other organisms had, and determine whether those fitness interactions were stable enough and held between enough organisms in order for evolution by natural selection to occur in virtue of those fitness interactions.[120] In the case of identifying a community as teleologically organized, it isn't enough that there has been at one time or other a trait-community or even such a community that was fitter than another; but

---

120. Glymour (2017) has argued these epistemic problems are largely intractable, at least by the current statistical tools and models we have for detecting selection.

there must be a lineage of such communities. These are extremely difficult epistemic challenges.

The epistemic hurdles don't undermine multilevel realism or even necessarily defining groups in terms of correlated interactions, but they make it extremely difficult to identify exactly which collectives are teleologically organized and so leave the proponent of an etiological account of teleological welfare in a state of uncertainty about which collectives have a welfare. And, from the moral point of view, even if we ultimately come to believe that teleological welfare is sufficient for moral considerability, the problem of epistemic inaccessibility means that we just won't be able to identify how the moral considerability of collectives figures into moral deliberations. Surely, on a plausible theory of what excuses wrong action, we must be excused from acting wrongly when we could not possibly ascertain the interests that would have grounded our acting wrongly. So perhaps biocentrism survives in practice, on this view, because individual organisms are the only bearers of moral considerability that we can really take into account.

Darwinian Individuals

In *Darwinian Populations and Natural Selection* Peter Godfrey-Smith provides an alternative to the trait-group framework and definitions of groups in terms of correlated interactions. Godfrey-Smith is interested in identifying not only the conditions under which natural selection operates, but those under which natural selection is likely to result in adaptation. His project starts by characterizing what he calls a "Darwinian population" and distinguishing two senses of "Darwinian population." "A *Darwinian population in the minimal sense* is a collection of individual things in which there is variation in character, which leads to differences in reproductive output (differences in how much or how quickly individuals reproduce), and which is inherited to some extent" (Godfrey-Smith 2009, 39). Every member of such a population is a "Darwinian individual" (Godfrey-Smith 2009, 40). Darwinian individuals need not be individual organisms. A Darwinian population can be a population of groups of organisms so long as those groups satisfy the relatively permissive conditions for counting as a Darwinian population.

A Darwinian population, in the minimal sense, is roughly just a way of describing a population that satisfies the Lewontin conditions. What distinguishes a minimal Darwinian population from a paradigm Darwinian population is that the latter has additional features that make those populations *evolvable*; these populations have features that allow natural selection to generate adaptations—to evolve novelty (Godfrey-Smith 2009, 41).[121] While Godfrey-Smith identifies a range of features as being relevant to evolvability, he defines a "paradigm Darwinian population" as one that exhibits a high degree of three specific features: fidelity of heredity, smoothness of fitness landscape, and dependence of reproductive differences on intrinsic character (Godfrey-Smith 2009, chap. 3). The first of these features refers to how closely offspring resemble parents. If inheritance is extremely imperfect, response to selection is diminished no matter how much fitter parent individuals are relative to other individuals in their generation. The second of these features refers to the degree to which variation corresponds to degree of fitness. Consider a fitness landscape that is very jagged—one on which there are high peaks surrounded by deep valleys. For example, we can assume that fitness varies with height such that individuals that are 6ft tall are just as fit as individuals that are 3ft tall, but any height between these is less fit and being 4.5ft tall corresponds to the lowest fitness. Let's assume that all organisms in a generation are roughly 6ft tall. In this example, it will be difficult for the population to evolve by natural selection to be 3ft tall. Natural selection will act to keep the population around the local fitness peak because slight variations in height are selectively disadvantageous. However, if the fitness landscape were smoother, for example, then individuals in the population might find themselves atop another nearby fitness hill with only a slight variation in height.

The last of these features refers to how much the fitness of individual is a function of intrinsic features of the individual. As an example, consider that a population of individuals is hit by a natural disaster which kills off a large

---

121. Godfrey-Smith (2009, 42–43) also articulates how it is that the Darwinian population approach to understanding evolution by natural selection figures into what he calls "origin explanations" of traits: explanations, in terms of selection, for why an individual bears a trait, or how it is that a trait came to exist at all, as opposed to just why it is that traits are distributed as they are within a population. There is a question, which I have set aside, about whether grounding teleological organization in natural selection requires that we be able to give origin explanations by appeal to natural selection or not. Sober (1995, 2014) has claimed that natural selection cannot furnish such explanations, while Neander (1995) has argued otherwise. I have set this aside because it seems to me orthogonal to issues of teleological organization. Whether or not natural selection can serve as part of origin explanations does not undermine, for example, Neander's etiological account of function or the use of etiology, as employed, in grounding claims of teleology. However, if Lewens (2004), also a skeptic about natural selection playing a role in origin explanations, is correct that sorting is sufficient for grounding teleology, then perhaps one might wish to adjudicate the debate about origin explanations and try to show that selection is required for origin explanations and selection must result in origin explanations in order to ground teleology.

portion of them. The survivors contribute more offspring to the next generation. But this higher fitness isn't due to intrinsic differences between those that survive and those that don't. To the extent that fitness differences are extrinsic, they are less likely to project into the future and so stick around long enough to result in adaptations.

On this framework an entity can be plausibly judged to be teleologically organized according to the etiological account of teleology only if it was part of a lineage of populations that approximate paradigm Darwinian populations. The features that make a population count as a paradigm Darwinian population come in degrees, but merely being a Darwinian individual in a minimal Darwinian population is not sufficient. To identify an organism or group as teleologically organized is not to say that it is subject to natural selection but that it has evolved by it, has parts and processes that are there in a current population because of the ends they brought about in an ancestral population.

There is no in-principle reason why biological collectives couldn't form a paradigm Darwinian population. Godfrey Smith acknowledges as much in his discussions of group selection (Godfrey-Smith 2009, chap. 6).[122] However, again, species and ecosystems are unlikely to fit the bill. First, it isn't clear that there is a population of ecosystems or species that would even count as a minimal Darwinian population. Second, even if there were, they would not be paradigm Darwinian populations simply in virtue of the fact that they have low fidelity of heritability. We need to understand not only species or communities as replicating themselves, as opposed to new generations just being byproducts of individual reproduction, but future generations as having to closely resemble their parent generation. In our toy example, groups of altruists don't create new groups of altruists but instead produce a certain number of individuals that are altruistic. In order to count as group selection, we would have to understand groups as reproducing so as to generate a future group rather than a collection of individuals who then pair up. This is an important difference between the Darwinian individual framework and the correlated interaction framework.

---

122. Godfrey-Smith mentions bee colonies as collections of organisms that he takes to satisfy at least the reproductive conditions he puts on Darwinian populations (2009, 119) and acknowledges that some individuals in tight symbiotic relationships, like lichens and the different types of individual bacteria that lead to the evolution of the eukaryotic cell, might also constitute Darwinian populations (Godfrey-Smith 2009, chap. 4).

How to Give Up Biocentrism

If I am right that multilevel realism is true, there is no clear answer to "Which collectives are teleologically organized?" However, given the possibilities across different realist conceptions of groups and group selection, it seems clear that the traditional wholes that holists might wish to license as morally considerable are not teleologically organized. Ecosystems (and species) traditionally understood don't result from collective sorting, don't constitute trait-groups or trait-communities, and aren't Darwinian individuals in paradigm Darwinian populations. So however biocentrists must expand their view about what has a welfare, they need not expand it that far. Beyond that, which collectives are teleologically organized becomes a matter of deciding between realist views, overcoming empirical challenges, gathering data, and developing models that would help us identify or predict where higher-level selection process might have been causally significant in the evolution of a trait.

However we ultimately decide these issues, biocentrism is false. Perhaps biocentrists will be willing to accept this and modify their view to accommodate the best views about the levels of selection. They might adopt a view I will later call teleocentrism, the view that all entities that are teleologically organized according to an etiological account of teleology are morally considerable. In what follows, I argue that this will require accepting more than just biological collectives, but also artifacts, and that teleocentrism is ultimately unacceptable.

THE STRATEGY OF EXCLUSION AND THE PROBLEM OF ARTIFACTS

Whereas biocentrists have typically seen the exclusion of collectives from the domain of moral considerability as simply the consequence of accounts of the welfare of nonsentient organisms and what they took to be the correct view about the level at which selection operates, they often take it as a condition of adequacy that artifacts, at least nonsentient artifacts, be excluded.[123] In other words, biocentrists might be willing to accept that, strictly speaking, their view is false and simply allow that some biological collectives will ultimately satisfy the conditions for being morally considerable, but they seem unwilling to accept that artifacts such as corkscrews, can openers, televisions, and typical computers are morally considerable, often taking it as a *reductio* of any view of moral status that it recognizes artifacts as morally considerable (Cahen 2002, 117; Sterba 1998, 371).[124]

In my view, we must recognize that artifacts have a welfare in the same sense that nonsentient organisms do. This is not a *reductio* of an etiological account of teleological welfare but simply a consequence of the most plausible view we have of the (true) claim that nonsentient organisms have a welfare. However, it does raise trouble for biocentrism. If biocentrists are to maintain their view and exclude artifacts from the domain of the morally considerable, it will have to be because the welfare of artifacts is not

---

123. Many of the arguments and ideas of this chapter are drawn from work I have done with Ronald Sandler and that has been published in article and book chapter form (Basl and Sandler 2013a, 2013b). I draw heavily on this work here.

124. Interestingly, Sumner, who thinks that teleological accounts of welfare and objective-list views more generally are implausible and cannot meet the requirement of subject-relativity, considers the fact that such accounts might license ascriptions of welfare to artifacts or even inanimate objects to be "mere embarrassments" for such theories (1995, 788).

morally significant. In the next chapter I make the case that there is no room for an asymmetry of moral considerability between nonsentient artifacts and organisms. In this chapter, I will make the case that artifacts have a welfare in the same sense that nonsentient organisms do. To do so, I'll first lay out the prima facie case that the etiological account of teleological welfare implies that artifacts have a welfare and point out what more will be required to deny that artifacts have a welfare. I then turn to the distinctions and conceptual resources that are most commonly appealed to in denying that artifacts have a welfare and show that they are inadequate, that, for example, appeals to natural vs. artificial selection or the derivative nature of artifact teleology cannot serve to exclude artifacts from being morally considerable.

The Prima Facie Case for Artifact Welfare

It is undeniable that artifacts, at least most of them, are teleologically organized; they are oriented toward achieving particular outcomes or realizing certain goal-states.[125] Some artifacts are organized toward a narrow range of ends (the purpose of a can opener is to open cans—the purpose of a corkscrew is to remove corks), while others, such as computers and their software, have a much broader range of ends. In the case of artifacts there is no prima facie mystery about how they could be teleologically organized. While it is fair to wonder whether organisms, absent a God that designed them, merely give off the appearance of teleological organization, it seems that the source of artifact teleology, some conjunction of our intentions and actions in realizing them, is readily at hand.[126]

Since artifacts are teleologically organized, we *can* define their welfare in precisely the same way that we define the welfare of nonsentient organisms. While we don't often talk about the welfare of artifacts using that language, we do often talk about what is good or bad for them, and this is most typically understood in terms of their teleological organization. Using a knife on a marble countertop is bad for it because doing so dulls its edge, diminishing its ability to realize the end of cutting; letting a car rust out or letting the tires go bald is bad for the car because doing so diminishes its capacity to realize the end of transportation. And while artifact ends almost always coincide with our ends,

---

125. There are some artifacts which are not teleologically organized. A pile of trash, thrown together at random, is not itself teleologically organized.

126. It is another issue to identify exactly how intentions and actions must come together in order to generate teleology. I set the details aside, but see Griffiths 1993; Lewens 2004.

they can come apart. Car collectors and knife enthusiasts might recognize that the current user of a car aims to ruin its tires or dull a knife's edge while they still see this as bad for the artifact.

The Generalized Etiological Account of Teleological Welfare

It isn't simply that we can make sense of or sometimes do talk about the good of artifacts. The account of welfare of nonsentient organisms already developed easily extends to ground claims about artifact welfare.[127] According to the basic etiological account of teleological welfare, something is good for nonsentient organisms if and only if it promotes one of that organism's ends (and bad for the organism if it frustrates some end). And an organism has ends in virtue of being a unit of selection. It is easy to see how this account could be generalized to include artifacts:

> **The Generalized Etiological Account of Teleological Welfare**
> Something, X, is good for a nonsentient thing, T, if and only if it promotes one of T's ends. X is bad for T if and only if it frustrates one of T's ends.
>
> T has E as an end if and only if there was selection for E at the level of T.

This account is structurally similar to the basic account: welfare is defined in terms of goals, and the unit of teleological organization, that which has the ends or goals, is understood in terms of selection at a level. As with organisms, there is a need to make reference to the level at which selection occurs. A pile of discarded artifacts is not itself teleologically organized even though every individual artifact that composes it might be. The reason for this is that there was not selection at the level of the collection; the agents that created the pile of trash, we can assume, had no aims or desires and took no actions to organize the collection of artifacts toward ends. Artifacts are teleologically organized, when they are, because they are the unit or target of selection. The reference to ancestral populations in the analysis or definition of having an end has been dropped simply because artificial selection doesn't always require iterative design and so there need not be an ancestral population.

---

127. See Griffiths 1993 for a discussion of the ways in which teleology across these contexts is unified.

If we accept the generalized account, then artifacts have a welfare—one definable in terms of what promotes or frustrates their ends. Though I haven't specified exactly how we are to understand selection in artifacts or made clear under exactly what circumstances or what combination of intention, action, or use one can claim that some artifact has been selected for certain ends, selection of some form must explain the obvious fact stated above: that artifacts are teleologically organized. It is in virtue of selection etiologies that we can distinguish things an artifact does from things it is supposed to do or is oriented toward. It is by appeal to selection etiologies that we can explain how it is that malfunctioning artifacts that don't achieve their ends still have certain outcomes as their ends. And it is because of selection etiologies that we can explain the relationships that hold between parts and wholes (such as in the case of the pile of trash) as well as when we wish to explain why the parts of a watch have purposes relative to the purpose of the watch as a whole. Once we have that artifacts are teleologically organized, we can define their interests or good in terms of their ends.

The biocentrist (and other proponents of an etiological account of teleological welfare) will surely insist that we should not move from the basic account to the generalized account, that being teleologically organized is not sufficient for having the sorts of ends that give rise to a welfare, or that being selected for isn't sufficient for the kind of teleological organization, genuine teleological organization, that gives rise to ends. They might claim that in the case of artifacts, it is not the artifact itself that has the end and try to explain why the selection etiologies of organisms and artifacts explain this difference. There are many avenues, to be considered below, for attempting to draw distinctions between organisms and artifacts to resist the move from the basic account to the generalized account.

Any conceptual resources deployed by biocentrists to resist the generalized account must satisfy three requirements. First, there is a requirement that the proposed difference between artifacts and organisms not be ad hoc or question-begging. It won't suffice to just say that genuine teleological organization requires that a thing belong to a biological species. This does the job in one sense but is entirely ad hoc. Second, it must turn out that all typical, living, nonsentient organisms have a welfare. Biocentrists can set aside instant organisms and ignore nonliving organisms, but it is simply part of their view that all naturally evolved, living organisms on Planet Earth are morally considerable and so must have a welfare. If a distinction or other resource used to exclude artifacts from having a welfare ends up also excluding a range of these typical organisms, then that distinction or resource will be shown to be insufficient. Third, it must turn out that all, or almost all, artifacts end up lacking a welfare. While biocentrists can and should admit that something like a conscious artificial intelligence with significant cognitive capacities will have a welfare, however they distinguish organisms from typical artifacts, it must turn out that all corkscrews, can openers, thermostats, computers, and software programs lack such a welfare.

Anyone who wishes to deny that artifacts have a welfare in the same way that nonsentient organisms do must explain what distinguishes these classes of things in a way that satisfies the above conditions of adequacy. Failing that, they must either accept that artifacts have a welfare or give up on claims that nonsentient organisms have a welfare. I've already argued that we have good reason to accept that nonsentient organisms have a welfare so long as we can provide an account of that welfare that meets the subjectivist challenge and that the etiological account of teleological welfare is the best and only game in town for meeting that challenge and grounding nonsentient welfare. In chapter 6, I'll reconsider what the balance of arguments tells us about where we should end up on questions of moral status and welfare, but, setting that aside, there is at least a burden faced by the biocentrist: given that the feature of nonsentient organisms that best explains or grounds their welfare is shared by many artifacts, why should we accept a restriction on the generalized account absent some compelling argument for that restriction?

The Argument from Synthetic Biology

One potent argument that cements that this is a genuine burden, an argument for accepting the generalized account and rejecting the salience of most of the purported differences between organisms and artifacts, comes from synthetic biology. "Synthetic biology" is a term used to encompass a variety of (sometimes loosely related) approaches to engineering biological systems from the ground up. To understand what is distinctive about synthetic biology, it is useful to contrast it with traditional genetic engineering. Traditional genetic engineering, including cloning, involves an extant genome derived from the genome of some naturally evolved organism. In the case of cloning, a copy of an organism's genome is made using analogues of reproduction. In the case of gene therapy or genetic engineering, a variety of techniques are employed to remove, replace, or add genetic sequences to the genome of an extant organism.

The tools of genetic engineering allow for radical modification of the genome. Take CRISPR for example. CRISPR is a technology derived from naturally occurring bacterial immune systems. What CRISPR does, very roughly, is seek out particular genetic sequences, unwind the DNA strand where they occur, and then cut out or replace those sequences. What is distinctive about CRISPR and why it has so much potential is that it does not simply do this in a single cell within an organism or embryo; it can be used to target many cells, for example all the cells in the germ line (Ma et al. 2017). So, for example, CRISPR could be used to ensure that deleterious genes are not passed on to future offspring because all such genes have been repaired by the CRISPR system. CRISPR can also be used to design genomes, perhaps even novel ones, by inserting sequences that don't exist or removing sequences that are typical of certain genomes. For this reason, CRISPR can be used to efficiently build model genomes for study.

Another way that genetic engineering allows for the design of novel genomes is by using the tools of genetic engineering to create transgenic organisms or engineered chimeras. An engineered chimera is an organism whose genome has been modified so as to include genes from organisms of at least two different species (Streiffer 2015). While many genomes might contain genetic sequences from other species, for example, those of viruses, genetic engineering allows for the insertion of a much wider range of genes from one species into another. Transgenic crops, such as Bt corn and golden rice, provide well-known examples. But there are also transgenic animals such as AquaAdvantage Salmon—salmon that have been engineered to reach their full size in half the time of naturally occurring salmon by having a genome modified with genes from another fish. There have already been attempts to insert human embryonic stem cells into mouse embryos in order to study human cell development (James et al. 2006). And work has been done to develop mice with components of human immune systems (Rämer et al. 2011), which, given the ways in which animals are often poor models for humans (Engels 2012), would be valuable for creating animal models that are better suited to testing, for example, drug effectiveness.

Synthetic biology departs from traditional genetic engineering in that the genomes and genetic sequences developed are built from the ground up.[128] While the tools of traditional genetic engineering can be used to develop novel genomes, genomes that do not and perhaps cannot occur naturally, the bits and pieces are drawn from organisms and trace back to organisms. Synthetic

---

128. Holm (2012) draws a distinction *within* synthetic biology between top-down and bottom-up approaches on which synthetic genomics is top-down and protocell research is bottom-up. This is not the distinction between top-down and bottom-up to which I'm referring.

biology does not take extant genes and put them together; it has the potential to build the genes themselves or even to create organisms from parts that have never occurred on earth. As an example, consider "synthetic genomics," a research program within synthetic biology. The goal of synthetic genomics is to synthesize DNA sequences, rather than borrowing sequences that already exist in an organism, embryo, or cell, and then use these sequences to create a genome and ultimately an organism (Kaebnick and Murray 2013, introduction; Holm 2012). There has already been substantial progress in the creation of organisms via synthetic genomics. In 2010, scientists from the J. Craig Venter Institute published a paper in which they describe the creation, from scratch, of a bacterial genome (Gibson et al. 2010). In this instance, researchers reconstructed the DNA sequences of an extant bacteria, so they built from the bottom up a genome that has the same sequence as a naturally occurring one. But there is no requirement that this be so. In principle, organisms could be built with novel genomic sequences. Or consider the development of "protocells," another research program within synthetic biology that aims at the construction of cells from components that do not occur in naturally occurring organisms (Hanczyc 2011; Holm 2012; Kaebnick and Murray 2013).

What is so distinctive about the products of synthetic biology is how they straddle the categories of organism and artifact, of natural and artificial. Humans have enmeshed these categories to some degree for a long time. Through hybridization and selective breeding, humans have taken what nature has furnished and modified it in varying degrees such that today's crops, agricultural animals, and pets are to a high degree artifactual. Genetic engineering has increased the extent to which we can impose our intentions and designs on organisms. Still, nothing humans have produced seems to be so firmly both an artifact and an organism as the products, or future products, of synthetic biology. Whereas genetic manipulation old and new resulted in organisms that were to some degree artifactual, the organisms still had links to the past that one could hope to appeal to in order to distinguish them from ordinary artifacts, the products of synthetic biology—call them *arganisms*—complicate matters significantly. Consider, for example, that on Varner's account, even the results of radical genetic engineering will still have sequences of DNA that can be ascribed a biological function on the basis of their selection history. Given this, such an organism will have biological interests and, thereby, a welfare in virtue of those sequences. In the case of synthetic biology, where each sequence is itself designed, no part has a biological function and so generates no biological interests on this account.

The fact of arganisms anchors a prima facie case for accepting the generalized account and thus artifact welfare. Consider two entities that are intrinsically identical except that one is an organism and one is an arganism. For the purposes of this example let the target organism, the one that will be duplicated in arganism form, be a fern. We can further imagine that the creators of the arganism intended to create it with the same ends of the target organism, so it is oriented toward the same things: its leaves soak up the sun to produce energy, its roots spread to soak up water, it reproduces by spores, etc. The fern, as has been claimed throughout, has a welfare. What about the arganismic fern copy?[129] It seems clear to me that it does have a welfare and that a substantial defense must be given if one is to reject the symmetry of welfare between the arganism and the target organism, just as an argument must be given for understanding instant (nonsentient) organisms as lacking a welfare.

The symmetry between arganism and organism cannot be accommodated by the basic account; the arganism will not have a welfare on this account. This motivates a move to the generalized account. Furthermore, the symmetry seems to immediately undermine many of the commonly cited differences between organism and artifact. The arganism is the product of design, of artificial selection; it is designed, ultimately, to satisfy some human end, in this case perhaps scientific curiosity, etc.

That the possibility of arganisms raises distinctive difficulties for articulating how it is that organisms have a welfare while artifacts do not has not gone unnoticed, both preemptively (Varner 1998; Jamieson 2008, 148) and concurrently (Attfield 2012a, 2012b; Preston 2008; Basl and Sandler 2013a, 2013b; Holm 2012). And while there are various moves one might make to untangle organism and arganisms or to accommodate their symmetry while denying that we must accept the generalized account and thus accept artifact welfare more generally, the burden rests squarely with those who wish to deny the generalized account to not only argue for some difference between organisms and artifacts that satisfies the conditions of adequacy above but also carves out space for arganisms or convinces us that we must reject the symmetry of arganisms and organisms.

---

129. I call it an "arganismic fern copy" instead of just "arganismic fern" because on some notions of species, this will not be a fern; it will have a different evolutionary lineage, which is enough on such views to count as a different species.

The Irrelevance of Purported Differences
between Artifacts and Organisms

While biocentrists and others opposed to recognizing that artifacts have a welfare might admit that they must be able to distinguish artifacts from organisms in a way that meets the conditions of adequacy described above, most have thought themselves to have a solution at hand. I consider some of the most obvious and most prominent purported solutions and show that each fails to meet at least one condition of adequacy.

Before proceeding, two notes. First, for any distinction discussed, the distinction may be taken as a way of amending either of the two components of the basic account so as to prevent a move to the generalized. So if the relevant distinction between artifacts and organisms that prevents an artifact's having a welfare is the living/nonliving distinction, it might be that being alive is necessary for predicating ends to a thing, *or* it might be that artifacts really do have their ends as their own but that this isn't sufficient for having a welfare, i.e., being alive is also necessary for having ends that ground a welfare. Some will have a problem with describing artifacts as having their own ends and some will be comfortable with that but think that we must therefore add additional conditions for having a welfare. In what follows, I will not always distinguish between these strategies in outlining the distinctions or in undermining their viability. I will instead frame how each distinction is taken to undermine the transition from the basic account to the generalized account in whatever way seems most natural to me, but the objections (and replies) don't hang on understanding them as challenges to one component or the other of the etiological account of teleological welfare. Readers may translate the objection into their favored language.

Second, this discussion cannot be exhaustive. There are innumerable surface-level differences between artifacts and organisms. For example, there were no human-made artifacts prior to about 200,000 years ago. So one could claim that it is not teleological organization that is sufficient for having an end, but being teleologically organized and having been the result of a selection etiology that started prior to 200,000 years ago. And, if for some reason this doesn't distinguish artifacts from organisms cleanly, I have no doubt that some conjunction of properties could be cobbled together to mutually exclude artifacts and organisms. But these approaches are clearly ad hoc; they simply assume that there is a relevant distinction between artifacts and organisms and then draw on any resources available to distinguish them without any principled defense of the approach. This is not only implausible but surely reopens the issue of whether the resulting account of nonsentient welfare meets the subjectivist challenge since whether a nonsentient thing has a welfare becomes, pretty obviously, arbitrary.

Living versus Nonliving

Many environmental ethicists, environmental activists, and others think that there is something special about living things—something that makes them importantly different than nonliving things, artifactual or otherwise. All of the distinctions to be discussed are attempts to explain this difference, to identify what it is about living things that is lacking in artifacts, that gives rise to their having a welfare. But one option is to assert that it is simply being alive (or alternatively being an organism) that does the relevant work. Organisms have a welfare because they are *living things* organized toward ends, and artifacts, organized toward ends or otherwise, are not.[130]

The living/nonliving distinction does a good job, prima facie, in terms of mutually excluding artifacts and organisms. Until humans started literally building organisms, there were no living artifacts, and all living things were, well, alive. And, if the living/nonliving distinction is the relevant one, it is easy to see how to maintain symmetry between organisms and arganisms; arganisms are alive and so satisfy the additional requirement for having a welfare.

Unfortunately, the living/nonliving distinction or organism/nonorganism distinction does a poor job of meeting the conditions of adequacy. First, there are several ways in which it is ad hoc. Consider the dialectic between sentientists, who have denied that nonsentient organisms have a welfare, and biocentrists. The former claim that there is no plausible account of nonsentient welfare. They ask, in essence, "Why should we think being alive is sufficient for or a plausible basis for having a welfare?" This objection, I have argued, is best understood as a requirement to meet the subjectivist challenge. To meet that challenge, I have also argued, the biocentrist should appeal to teleological organization and ground that teleological organization etiologically. This provides an answer to what it is about living things that grounds their having a welfare. But then the objection is raised that artifacts are teleologically organized. Why shouldn't we view artifacts as having a welfare in the same way? What is special about being alive? Why is being alive an additional

---

130. Attfield at times comes pretty close to this view. His account of welfare is one on which the good of a thing is defined by species-typical or species-essential features (Attfield 1981, 42), but many artifacts obviously are members of some kind and so can be said to be kind-typical or not and have their welfare defined that way.

requirement of having a welfare? To assert that the living/nonliving distinction is the relevant difference doesn't help to answer any of these questions; it simply draws the boundaries in the way that biocentrist would like them drawn rather than providing an argument that the property of being alive is somehow essential to having a welfare. The organism/nonorganism distinction fares no better if "organism" is just a way of picking out living things.

One might argue that it is perfectly legitimate to appeal to a distinction precisely because it draws the boundaries in the most plausible way. Even if the living/nonliving distinction can't be given any further defense, the fact that it best accommodates our intuitions about clear cases is reason enough to accept it as the relevant distinction. As will be discussed in chapter 6, however we ultimately resolve the debate over the moral status of nonsentient organisms, there will be untoward consequences. Perhaps the best option is to accept the living/nonliving distinction as the relevant difference between organisms and artifacts that explains why the former but not the latter have a welfare.

This response is fine as far as it goes. The fact that a distinction accommodates what are taken as known data points or givens might be a pro tanto reason to accept it, or, at the very least, not reject it. But if biocentrists are to give a satisfactory answer to the subjectivist challenge, they must at least spell out what it means to be "alive" or be "an organism" in a way that is not just pointing, arbitrarily, to the things they would like included among the morally considerable. But what characterizes "life" or "being alive" or even "being an organism" is a difficult problem (Agar 1997), and there are various approaches to understanding these notions. Further reflection on the living/nonliving distinction shows that it is either unacceptably ad hoc and cannot meet the subjectivist challenge or else fails to meet the requirements for plausibly distinguishing organisms from artifacts, for preventing a slide from the basic to the generalized account.

One approach to spelling out the living/nonliving distinction or the organism/nonorganism distinction is by appeal to the stuff that these respective things are made of—that the building blocks of living things are what sets them apart. However, this approach seems doomed. Imagine that an engineer were to build, completely out of (dead) wood, a clever machine that adds layers to itself, seeks out more wood, and creates copies of itself. Is such a machine made of the relevant stuff to count as being alive? Presumably the answer will have to be no, but it is made of exactly the same stuff as something that is alive. It won't do to insist that the difference is that a tree is alive while the wooden machine is not; that is exactly the distinction to be explicated. Second, imagine that on some alien planet there has evolved by natural selection some form of life that is made of stuff identical to some earth machines or some potential earth machines. The Transformers, fictional sentient robots, might be like this; in the fictional universe of which they are a part, they are an alien species and, we can assume, do not have a creator, but evolved by natural selection. If there are robotic plants on the Transformer planet, do they count as being alive? Do they have a welfare in the same sense as our organisms? To answer no is to adopt an extremely anthropocentric or Earth-centric attitude about which kinds of building blocks could give rise to nonsentient welfare. But, if the answer is yes, and yet our typical artifacts can't possibly be made of the right stuff, then it is unclear exactly how "the right stuff" is being picked out except by pointing at it. Surely, the fact that we can only point to the things we want to count as having a welfare is unacceptably ad hoc. Finally, the building blocks approach can't even be used to distinguish living from nonliving/dead organisms since they are made of the same stuff. For all these reasons, this approach fails.

An alternative approach is "capacitative." On this approach living systems are characterized by certain capacities such as a capacity for growth, reproduction, or self-maintenance, or by having capacities that make them dynamic systems able to interact with the world rather than merely be acted on. Most of the things that typically pass as organisms or being alive satisfy capacitative criteria like those just outlined. Nonsentient organisms, from single-celled bacteria to plants, are internally organized so as to be responsive to external stimuli; they grow and self-maintain, engage in activity that contributes to that growth and self-maintenance; they metabolize, drawing resources from their environment in ways that further their ends. There are notable edge cases, such as viruses, that depend on other living systems to carry out certain functions, such as reproduction, and so whether these things count as being alive or as organisms will depend on how we fill out the capacitative approach. This is no problem for the biocentrist or anyone who wishes to draw lines between artifacts and organisms so long as we accept that some things that are made up of stuff similar to living organisms just turn out not to be living.

While a capacitative approach might be sufficiently inclusive in the sense that it captures all or most all organisms (and arganisms), this approach fares badly in satisfying another requirement: excluding artifacts. Choose whatever capacities you like and there are many artifacts that have them. Thermostats, or heating and cooling systems more broadly, are self-maintaining systems that take actions to maintain the conditions for their own existence. The temperature in a room drops, a thermostat registers this and sends a signal to a controller requesting more heat, the controller sends a signal that fires up a burner and turns on a fan, warm air is sent to a particular room, the thermostat registers this and sends a signal to the controller to turn off the burner and fan. If the failure of the system is sufficiently large, the whole system might be discarded, burn itself out, or otherwise break. Or consider computer systems and their software that are built with many mechanisms for self-maintenance. To prevent malfunction and avoid malicious software, a computer will search for updates, automatically apply patches, and take steps to remedy or remove malicious code. This is true not just of supercomputers or artificially intelligent systems, but of mundane personal computers. Or consider software programs and games that have characters that grow and develop in response to user input or in response to features of the virtual world. Or software that creates software or reproduces itself.

One might object that while some, even many, artifacts might have the capacity for maintenance, they don't have the capacity for *self*-maintenance. Organisms' behaviors or actions are directed at self-preservation, or organisms have the tools to self-maintain internal to themselves, whereas artifacts might maintain themselves, but this is not what their behaviors are directed toward, or the resources for maintenance are not internal to the thing itself. The distinction between self-maintenance and mere maintenance echoes how Aristotle understood the distinction between the natural and the artificial (Preston 2008; Aristotle 1999). It is also a way of cashing out autopoiesis and helps us to understand how it is that living organisms are autopoietic systems. Proponents of autopoietic accounts of teleology or welfare might find this to be the relevant difference between artifacts and organisms even if they agree with the conclusions in chapter 3 that autopoiesis doesn't ground teleology and that selection etiologies are essential to having ends.

How exactly is this distinction between maintenance and self-maintenance to be characterized? One type of view, call it the historical view, is that whether an action or behavior is self-maintaining depends on some historical fact about the action or behavior. Does it exist to serve the purpose of another being? Does explaining its existence require that we cite the purposes or ends of another entity? These questions depend on answers about the history of the thing in question. I consider such views in depth below, but at least some proponents of autopoietic or self-organizational accounts think that the relevant properties are not grounded in such historical facts, but rather in the way organisms and artifacts behave, act, and interact with their environments. This type of view is ahistorical.

There are various ways one might spell out the relevant causal interactions. One option, offered by Holm, who draws on organizational accounts of function, characterizes self-maintenance "as a property of systems that are able to exert a causal influence on their surroundings in order to maintain (at least some of) the boundary conditions required for their own existence" (2012, 537). On this view, self-organization is understood in terms of causes being self-directed. Another option would be to understand self-maintenance as a kind of independence; a system is self-maintaining when or to the extent that it doesn't require assistance from external sources. A third option is to point to mechanisms that ultimately generate behavior and note whether they are internal or external. Some like Rolston (2003, 145) have been impressed with the way that the behaviors of organisms are coded into the DNA that resides in every cell of an organism.

None of these ahistorical approaches will meet the requirements previously outlined. If we understand self-maintenance or autopoiesis in terms of being internally organized to maintain the conditions for one's existence, the objection that artifacts do not engage in self-maintenance fails. Many of the artifacts described above do things that contribute to their continued existence. If my heating system does not provide heat, it will be removed; if my antivirus fails, my operating system might become so corrupt as to no longer function. Even worse, as Holm notes (2012, 537), and as was discussed in chapter 3, there are things that are neither organisms nor artifacts, such as candle flames, that satisfy this condition for self-maintenance.

Nor does it help to try to make the distinction in terms of the degree to which a thing is self-maintaining by understanding the degree of self-maintenance in terms of how many behaviors are involved in maintenance. Even setting aside the challenge of specifying how many behaviors have to be directed at maintenance to count as self-maintaining, there are artifacts that exercise a huge number of behaviors that can be described as causally contributing to their own existence, especially in the case of very sensitive applications. The software/hardware systems that constitute (current or future) self-driving cars, systems used for space travel, for military applications, etc., perform a huge number of functions, many of which are required for the system to function at all and certainly in order not to be discarded for a better alternative. However much the simplest organism does to contribute to its own existence, imagine a very demanding set of computer users who share a computer and will destroy any computer that fails to perform their desired functions. Is it possible that the demands of this group are such that the computer performs as many behaviors as the simple organism does to prevent its destruction? It is possible. This example seems clearly to undermine any attempt to understand self-maintenance in terms of having some high degree of causal contribution to one's own existence.

What about understanding self-maintenance in terms of independence? Artifacts do depend on circumstances outside themselves—for example, they may require a power source, user input, or other resources in order to exercise or realize their capacities—but this is certainly true of organisms. Organisms, like artifacts, vary in the degree to which they depend on their environment in order to exercise their capacities, but some organisms are very dependent on their environment to exercise many of their capacities. Many animals depend on their parents' supplying resources, whether it be a bird that warms an egg or a mammal that provides milk to an infant, in order to develop. Organisms that have a way of life that is heavily symbiotic or parasitic will depend, in some circumstances, on other organisms to exercise even the capacities for growth and reproduction. Consider that we would not survive without a whole host of bacterial species that reside within us and contribute to metabolism (Gill et al. 2006). Or take the example of *Ophiocordyceps unilateralis*, or zombie ant fungus, which infects the brain of an ant, influencing the ant's behavior so as to enable it to infect more ants, continuing its life cycle (Evans, Elliot, and Hughes 2011). Or consider carnivorous plants of the genus *Byblis*. The species within this genus are "pyrophillic," meaning that their reproductive cycle is triggered by or otherwise depends on the presence of fire. In the case of members of *Byblis*, there is a chemical in smoke that triggers seed germination and without which germination would not occur or would occur much more slowly (Cross et al. 2013).

And there are artifacts that are fairly independent in the relevant sense. Take, for example, the Beverly clock. Invented in 1864 by Arthur Beverly, the Beverly clock is an extremely efficient mechanism powered by changes in atmospheric pressure. It has been running, with few interruptions, since its invention (Amon, Beverly, and Dodd 1984). Yes, the clock has stopped several times, due to its requiring maintenance or too few fluctuations in atmospheric pressure (Amon, Beverly, and Dodd 1984), but it isn't hard to imagine an improved design that doesn't require such maintenance. We can also imagine solar-powered computers that operate with fair amounts of independence. Whether an improved Beverly clock or a solar-powered computer that requires little maintenance, these satisfy the conditions for self-maintenance on the independence view.

As a final ahistorical view, let's consider the internal mechanism view, the view on which a thing is self-maintaining when maintenance behaviors or actions are directed from some mechanism internal to that thing. In the case of organisms, the thought is that DNA sequences (the genome), in an important sense, direct or code for those organisms behaviors and actions, and that is why these behaviors or actions count as self-maintaining.

While some have thought there is something to the idea that DNA or the genome is distinctively internal in a way that marks the distinction between the way artifacts behave and the way organisms behave, it is hard to see how exactly this is supposed to work. Is it that DNA is internal to the organism? This can't be it. Many artifacts have their behaviors directed fully by some internal mechanism. Rolston characterizes the genetic code as a "linguistic molecule" and as a "propositional set" that "Given a chance . . . seek organic self expression. An organism, unlike an inert rock, claims the environment as source and sink, from which to abstract energy and materials and into which to excrete them. It 'takes advantage' of its environment" (Rolston 2003, 145). Computer programs again seem especially good analogues of DNA. Consider, for example, a computer system containing a variety of software. A sequence of code runs, instructing the system of which it is a part to perform certain operations. And in the same way that DNA or the genome is an internal source of the behaviors that end up maintaining the system, those operations generate the conditions for their own existence in many senses. If the code were buggy or couldn't control the hardware, it would be discarded. The operations performed to prevent viruses, maintain system functions, etc. originate internally from the code, and all those behaviors or functions emanate from within the system, directed by the code. Software, too, is a propositional set.

Is it instead that the genetic code is somehow taken to operate independently of external input such that even though organisms might depend on external resources, they are generated from a source that is independent in some relevant sense? The view that the genome is just a sequence of code that executes independently of environmental factors and directs the development and behavior of an organism is outdated and mistaken. Epigenetic, or genic environmental, factors influence which genes are expressed and when, and the genic environment can be influenced by environmental factors. It isn't just that which behaviors are expressed depends on the environment, but which genes are expressed. What this means is that DNA is sensitive to the environment in a way that is similar to some software, its operation and outputs depending on external outputs.

Even if it were true, which it is not, that the genetic code of organisms was in some sense fully internal, insensitive to the environment, or at least didn't require external inputs to execute, this version of self-maintenance again fails to exclude artifacts. Unless one insists that only DNA provides the kind of internal mechanism for generating behaviors or actions that could count as self-maintaining (another version of the building-blocks approach already rejected), then there are clear cases of analogues that execute their own code and where that code enables or results in the performance of behaviors that are self-maintaining.

Natural versus Artificial

Several environmental ethicists have taken the fact that organisms are natural while artifacts are nonnatural as the central difference between artifacts and organisms. Naturalness on most construals, unlike those properties so far considered, is a historical property; two entities might be intrinsic duplicates and yet differ with respect to whether they are natural or nonnatural. This is not a reason to rule out naturalness as distinguishing artifacts from organisms. According to the etiological account of teleology, facts about origin, about selection history, are central to questions of welfare. A nonsentient organism cannot have a welfare if it lacks a certain kind of history precisely because lacking that history precludes it from being teleologically organized. The question is whether naturalness is a property like this, a historical property of a thing that is relevant to whether it has a welfare.

Naturalness has been cashed out in numerous ways, and there is a long history of trying to analyze "natural," disambiguate different senses of "natural," and identify the relevant contrast classes such as the supernatural or the artificial (Mill 2009). There are questions about whether naturalness is binary or, more plausibly, comes in degrees and of how much human intervention it takes to undermine naturalness. Probably every environmental ethics class spends some time articulating the challenges to understanding the natural and to finding any normative significance in the natural.

It is difficult to identify a sense of natural and its contrast class that would satisfy the necessary requirements. If being natural requires complete independence from human influence, then there is nothing that is natural anymore (McKibben 1999). How about distinguishing artifacts from organisms in terms of being "natural enough"—of having a sufficiently high degree of freedom from human influence? Whether this is a workable solution depends on there being some way of accounting for degrees of naturalness in more than an intuitive way, some way of identifying which sorts of human influences count as increasing unnaturalness and by how much. Absent that, it is hard to assess whether it can be used to cleanly, or cleanly enough, distinguish between organisms and artifacts.

The most plausible way of understanding naturalness, a way that, on its face, provides a non-ad hoc way of neatly dividing up organisms and artifacts, is by appeal to natural as opposed to artificial selection (Varner 1998; Preston 2008; Taylor 1989, 121–24). Let us grant that artifacts and organisms are teleologically organized. Even so, the source of that teleology, at least prima facie, differs in identifiable ways. In the case of artifacts, they result from selection etiologies that involve the intentional actions of human agents, whereas all organisms (setting aside arganisms), even those that have been subject to hybridization and selective breeding or cloning, have parts and processes that evolved by natural selection independent of our intentions or aims. By appealing to differences in selection etiology, we can distinguish natural teleology from artificial teleology and claim that natural teleology is necessary for a nonsentient thing to have a welfare.

This approach, at least on its face, has a lot to recommend it. First, it seems to do a nice job of neatly dividing up artifacts and organisms. Take paradigmatic artifacts such as corkscrews, can openers, computers, thermostats, and the like. The source of teleological organization in these cases is artificial selection. On the other hand, the source of teleology in "typical" organisms, setting aside for a moment GMOs, clones, arganisms and the like, is natural selection. This view even does a nice job handling difficult cases of organisms that have teleology derived from different sources. Take, for example, GMO crops like golden rice. Golden rice has been genetically modified so as to produce beta carotene and was developed to address vitamin A deficiency in humans (see Sandler 2007, chap. 6). While it is true that golden rice produces beta-carotene as a result of artificial selection and so has ends as a result of artificial selection, it grows, photosynthesizes, etc., as a result of natural selection etiologies. We can divide up the ends of genetically altered organisms into those that are natural and those that are artificial and understand only the natural ends as constituting or grounding welfare. This is especially useful in the case of organisms that have been modified in ways that seem detrimental to the organism, such as OncoMice.[131] OncoMice are mice genetically modified so that they will develop certain forms of cancer so as to be used in cancer research. In a sense, an OncoMouse is teleologically organized toward the end of getting cancer, but on the view under consideration we can avoid the implication that it is good for such a mouse to get cancer because this is not one of its natural ends in the sense of having been selected for by natural selection.[132]

A second virtue of this view is that there seems to be a principled reason for distinguishing the ends grounded in natural selection from the ends grounded in artificial selection. When a thing is teleologically organized because it has been artificially selected, it seems plausible to say that it is not

131. Delancey (2004) uses the example of OncoMice to generate an objection to Varner's account of biological welfare. However, his revision makes use of a nonetiological account of function, and so while it may generate plausible results in the case of OncoMice, it falls prey to the various concerns raised in chapter 3.

132. While useful in the case of OncoMice, the view does a poor job accommodating arganisms.

oriented toward its own ends, but oriented toward ours. To the extent that a thing is artifactual, it lacks a good of its own because it lacks ends that can be properly characterized as its own. On this view, it seems, there is not simply an insistence on the basic over the generalized account; rather the basic account is to be preferred because the generalized account fails to distinguish between sources of teleology that are relevant to having a good of one's own and thus fails to accommodate the derivative nature of artifacts and its implications for welfare.

The Derivative Nature of Artifacts

Despite its virtues, the view that a difference in selection etiology marks the difference in welfare between artifacts and organisms is false. By looking more carefully at how it is that artificial selection might be taken to undermine claims that an artifact's ends are its own, it becomes clear that artificial selection, like natural selection, generates genuine teleological organization, resulting in a thing that is organized toward satisfying its own ends, one that has a welfare definable in terms of those ends just as organisms do.

One way that artificial selection might be taken to give rise to ends that are derivative is that when something comes about by artificial selection (or in the case of things like GMOs, to the extent that they come about by artificial selection), there is no explaining the ends of that thing without reference to the ends of another being. Why is it that a thermostat regulates temperature? Because humans desire regular temperature and because some inventor(s) with some ends of their own thought of a clever tool for regulating temperature. Why is that antivirus software acts to remove malicious code from my laptop? Because I, like most computer users, want a computer that is virus-free, runs smoothly, and doesn't compromise my information, and some developers with ends of their own developed some relatively autonomous software to address my desires. For every artifact that is teleologically organized, it seems that a complete explanation for why it has the ends that it does references our intentions. The reason for not moving to the generalized account, of thinking that artificial selection etiologies generate ends that can't be used to define or constitute welfare, is that artificial selection histories make the ends of artifacts derivative in this explanatory sense, call it "explanatory derivativeness."[133]

---

133. This terminology is taken from Basl and Sandler 2013a. For biocentrists who seem to think this is the relevant distinction between organisms and artifacts see Goodpaster 1978, 319; Taylor 1989, 61.

There are several reasons for rejecting explanatory derivativeness as that which undermines the move to the generalized account. First, there are ends of organisms that both define, or are constitutive of, welfare and are explanatorily derivative. This isn't just true of GMOs, clones, and synthetic organisms, though it is true of those as well. Think of how the evolution of cognitive capacities in some species affected the fitness landscape for other organisms. Once there are organisms that have intentions and that act on those intentions, traits that didn't previously impact fitness might suddenly do so. As the number of creatures with intentions increases, more and more traits of other organisms evolve in a context where a complete explanation of those traits and what they are for will invoke intentions. If any of those traits serve ends which define or are constitutive of welfare, then explanatory derivativeness does not undermine genuine teleological organization.

As a first example of ends that are explanatorily derivative but still define the good of an organism, consider domesticated dogs. In the Canidae family, the domesticated dog's most recent common ancestor is the wolf (Bradshaw 2012). Humans began domesticating dogs perhaps as long as a 100,000 years ago (Bradshaw 2012). Since then, dogs have diverged from wolves in many ways, adapting to us and our needs (while we have also adapted to theirs). Two interesting differences between wolves and dogs are that dogs have evolved to expect or seek out help with difficult tasks and dogs are much more responsive to human gestures, requiring much less training than wolves to respond to gestural cues (Miklósi et al. 2003). Another important difference, due to dog-human coevolution, is that when dogs engage in mutual eye contact with humans, there is an increase in oxytocin levels in both the dog and the human that facilitates bonding (Nagasawa et al. 2015).

It is good for a dog to be responsive to gestures, to be able to seek out the aid of humans, and to have the mechanisms that aid in bonding. To undermine these abilities or capacities is to harm the dog, to reduce its welfare. This is so because the dog has cohabitation with humans as ends, because there are parts and processes that exist in dogs because they were selected for cohabitation and, ultimately, for survival and reproduction. And yet a full explanation of the traits of dogs requires an appeal to our intentions. Dogs have these traits, in part, because past dogs served some ends of ours, and thus these ends are explanatorily derivative. If these are genuine ends of dogs that constitute or define, at least partially, their welfare, then explanatory derivativeness is irrelevant to welfare.

It might be tempting to try to avoid the objection by pointing out that the reason cohabitation is good for dogs can be fully explained by appeal to their sentience and that, because the proposed ends are explanatorily derivative, we should reject that they are constitutive of welfare. But I urge the reader to avoid this temptation. The example of dogs is useful partly because dog-human coevolution has been the subject of serious scientific inquiry and so gives the flavor of the problem of trying to appeal to explanatory derivativeness without having to tell a "just-so story" about the evolution of organisms. But once we recognize the way that intentions can figure into the evolution of traits, it is easy to imagine that there are many nonsentient species whose ends are explanatorily derivative and yet still constitutive of their welfare.

There are some wild bottlenose dolphins that break off pieces of sponge to probe the ocean floor for food (Krützen et al. 2005). I take it that this use of tools, like tool use in other species such as nonhuman primates and some birds, is intentional. The dolphin case is particularly interesting partly because the tool in use is another organism, a nonsentient one, that can reproduce asexually by fragmentation and budding. The case is also interesting because it seems that this is an instance of tool use that is culturally, as opposed to ge-netically, transmitted (Krützen et al. 2005). The specific use of sponges as an ocean probe is passed on from generational to generation by demonstration or teaching.

With these facts as background, I will engage in some just-so storytelling. Imagine that dolphins' use of sponges has become an important part of the evolutionary trajectory of those sponges. Let's just say that dolphins have had a preference for sponges of a certain color, a certain size, and certain shape. Over time, sponges that better fit the properties desired by dolphins did better. After many generations, most of the sponges that exist in areas populated by dolphin tool-users are one foot-ish long, tubular in shape, and bright red. A bacterium comes along that alters the color, shape, and size of sponges, making them less likely to be used as tools by dolphins and so less successful at reproduction.

What should we say about the good of these sponges? It seems clear that the bacterium is bad for the sponges, it harms them. It does so because it undermines the realization of their biological ends: the ends of growing to a certain length, to a certain shape, of having a certain color and, ultimately, because it undermines the end of reproduction. But these ends are clearly ex-planatorily derivative; a full explanation of the ends of the organism will refer-ence the intentions of the dolphins. If an end's being explanatorily derivative undermines its grounding or defining welfare, then the sponge doesn't have interests that it clearly has.

Is it really true that we must reference the intentions of the dolphins in explaining why the sponges have the color, shape, and size that they do? Can't we simply explain things in terms of the differential rates of survival and reproduction of sponges of different colors, sizes, and shapes of sponges? There are some sponges that are long, some that are short, some tubular, some that are not, and some that are bright and some that are not. The reason there are now mostly red, footlong, tubular sponges is that sponges that lacked these properties failed to reproduce as well as those that had these properties. This is because something in the environment, a dolphin in this case, broke apart sponges that were tubular, red, and foot-longish, enabling greater rates of reproduction by fragmentation. This explanation doesn't make mention of the intentions of the dolphins, it just describes their actions.

This strategy of describing the selection history of the sponges purely in terms of the actions without reference to the intentions of dolphins seems perfectly fine. Someone who wanted to know why sponges in an environment were as they were might not care about the dolphins' intentions or ends, and so this explanation will be fully adequate for this explanatory purpose. But this strategy is available in the case of every artifact. We can explain artifacts' having the internal components and mechanisms that they do simply by explaining what those components and mechanisms do or what they do better than some alternative and without referencing the intentions of the designer. Why does my thermostat have buttons instead of a slider? Because sliders break more easily. Why does my computer auto-update? Because computers that didn't auto-update ended up becoming slow and were thrown out.

We can admit that different levels of explanation, different descriptions of, for example, causes, are appropriate in different contexts depending on our explanatory aims (Van Fraassen 1977, 1980), but this doesn't help to shore up the argument that the explanatory derivativeness of artifacts' ends undermines their having ends that are their own. It just provides a strategy for describing those ends in a way that is not explanatorily derivative. To whatever extent a full explanation of the ends of artifacts must reference intentions, it is hard to see why the same consideration wouldn't apply in cases like the sponge.

It also won't do to simply insist that a thing's ends are explanatorily derivative only if the relevant intentions referenced in explaining its ends are those of humans. For one thing, it would just seem odd to accept this kind of human specialness, this kind of anthropocentrism, in the context of arguments in environmental ethics where the aim is often to undermine human specialness. More importantly, though, *Homo sapiens* have walked the earth for approximately 250,000 years. Let's imagine that our early ancestors shaped the evolutionary trajectory of many nonsentient organisms in just the way we imagined the dolphins shaped the evolutionary trajectory of the sponges. We can imagine that many grasses, many wild fruits, many insects, etc. have the ends they do because, or partly because, of some intentions our ancestors had.

This hardly seems relevant to whether the ends of those organisms constitute the good of those organisms. It doesn't matter that this might not be true, that no organisms' ends really are explanatorily derivative in this way. What matters is that, counterfactually, explanatory derivativeness would not at all matter to whether these nonsentient organisms had a welfare definable in terms of their explanatorily derivative ends.

So far I've argued that the ends of a thing being explanatorily derivative does not undermine the claim that such a thing has those ends as its own by arguing that natural selection etiologies sometimes involve the intentions of organisms, i.e., that natural selection occurs via intentions and thus makes the ends of at least some organisms explanatorily derivative. A second reason to reject explanatory derivativeness as relevant to whether a thing has its ends as its own or whether those ends constitute welfare supposes that there are no natural selection etiologies at all. It is possible that Darwin was wrong and that creationists are correct—that God intentionally crafted every species of organism that exists on the planet. We can assume that God intended for every organism the same ends that it actually has via natural selection. So, for example, God intended that pine trees have needle-shaped leaves so they can photosynthesize through the winter and, ultimately, so they can survive and reproduce in the particular conditions they are likely to face. Their having these as ends is ultimately explained by God's intentions.

If creationism is true, do all nonsentient organisms cease to have a welfare? If the truth of creationism precludes providing an account of welfare that meets the subjectivist challenge, then the answer is yes. Just as we must accept that instant organisms lack a welfare once we see that there is no plausible account of their welfare, we would, in the face of the subjectivist challenge, have to reject all nonsentient welfare if creationism precluded the possibility of giving a nonarbitrary, nonderivative, subject-relative account of welfare. But the truth of creationism doesn't preclude this. To the contrary, in a creationist world, it is quite natural to understand the good of organisms in terms of their ends and to explain those ends in terms of the creator's intentions. It is once we recognize that we are in a Darwinian world that teleological theories of welfare face a challenge—a challenge that is met by recognizing that natural selection grounds teleology. Recognizing this doesn't give us any reason to suspect that when we explain why a thing has the ends that it does in terms of intentions that the thing ceases to have those ends as its own, that those ends aren't as constitutive of welfare as the ends of naturally evolved organisms.

There is another sense in which artificial selection might make the ends of organisms derivative and thus, not their own. While organisms are often used to satisfy our ends, artifacts exist solely for this purpose. It isn't simply that the ends of artifacts can't be explained without appeal to the intentions of others, but that their very existence depends on their being useful to us. The idea then is that artifacts, or things that are teleologically organized solely on the basis of artificial selection, do not have their ends as their own for the same reason that our fingers don't have ends of their own: they are merely parts or tools that are part of a larger teleological system and exist to help realize the ends of that system. Their ends are really our own (Basl and Sandler 2013b).

The fact that something only exists to serve our ends does not undermine its having its own ends. Elsewhere, Ron Sandler and I have used the example of a child that has been selected for its predisposition toward musicality (Basl and Sandler 2013a). We can imagine that a couple has taken the best available information about the genome and worked to identify genes or gene clusters that predispose their bearer to have highly sensitive ears, nimble fingers, a large vocal range, etc. We can imagine that the parents use preimplantation genetic diagnosis to screen for embryos that have the desired genes and implant an embryo only if it has all of them. The resulting child exists only to realize the ends of the parent; had the parent not wanted a musical child, that child never would have been implanted. And yet this doesn't undermine claims that the child is organized toward ends; it doesn't cast doubt on whether the child has ends of its own. If the child is encouraged, not manipulated or coerced, to take up music and does so, coming to love playing and listening to music, and ultimately taking it as an end to make music, music-making would be among the child's ends, and the child's flourishing might depend greatly on realizing that end even though the child would not exist if musicality of the child were not the end of its parents.

The same kind of reasoning applies when we discuss the selective breeding of organisms or the genetic modification of organisms. It isn't simply that we can't explain the ends of agricultural crops or domesticated animals without reference to our intentions; it's that most would not exist if they didn't serve our ends. This doesn't mean that they don't have their own ends, that because they exist only because they serve us that their ends are not their own.

And, again, this version of the argument doesn't survive the possibility that creationism is true. In the creationist scenario, everything ultimately might exist solely to serve the ends of the creator. In the fictional world of the *Hitchhiker's Guide to the Galaxy*, the earth is a computer populated as it is because an alien species is using the earth to determine the answer to a particular question.[134] The earth ends up being destroyed. Should we say that the destruction of the earth in this fictional world was not bad for the nonsentient

---

134. The question is "What is the question to which the answer is 42?" where "42" is the answer to "life, the universe, everything" (Adams 1995).

life on the planet? After all, that life existed only to serve the ends of another alien species.

Both these attempts to undermine the goal-directedness of artifacts make a similar mistake. They confuse the *source* of teleology and the *subject* of teleology (Basl and Sandler 2013a). There are many examples where we might appeal to historical facts, including facts about past intentions, to explain why a thing has some property or that it exists at all. This doesn't undermine *at all* claims that the thing is the bearer of that property. I am just barely over six feet tall. To explain that, one might invoke many facts about my family, including their intentions both in deciding to have a child and deciding what to feed me. This doesn't in any way undermine the fact that it is I that am six feet tall. The same can be said about being goal-directed: whether a thing is goal-directed or has its own goals is compatible with many different explanations for that goal-directedness and compatible with the fact that a thing might not have existed if it didn't serve the purposes of another being. This is so even though if some things had been different, the thing would not have been goal-directed or would not have existed.

The Case from AI and Machine Learning

Previously, I appealed to technological developments in synthetic biology to make the prima facie case for the generalized account. Here I'd like to consider another technology that raises difficulties for distinguishing artifacts from organisms: artificial intelligence. By "artificial intelligence" I do not intend conscious or sentient systems. Artificial intelligence is understood much more broadly to include many technologies which approximate intelligence or which solve problems in novel ways that they aren't specifically programmed to. Recent developments in machine learning allow for the possibility of dynamic, self-maintaining systems, capable of undirected, novel behavior. These systems and their behaviors can even come about via processes akin to (if they don't just count as) natural selection.

Traditional algorithms can be thought of as functions mapping input to output. Traditional algorithms can be very complex—for example, they might be able to play chess and beat all but the very best players—but there is still a directed mapping of input to output. The algorithm might be written such that if a user moves a white pawn from e2 to e4 on the first move of the game, it will move its pawn on e7 to e5. We can imagine writing an algorithm that covers every possible move in the game and has corresponding outputs. Programmers are very clever and so algorithms can be very clever. Still, there is a sense in which a traditional algorithm is just carrying out the very specific commands of its designers and it is limited by the rules that programmers are able to instantiate in code.

Machine learners are algorithms, too, but they differ from traditional algorithms in that they are not told which specific outputs correspond to which specific inputs. Instead, machine learners learn how to solve the problems they are tasked with. Here is how Pedro Domingos, an expert in machine learning, describes what is distinctive about machine learners:

> Every algorithm has an input and an output: the data goes into the computer, the algorithm does what it will with it and out comes the result. Machine learning turns this around: in goes the data and the desired result and out comes the algorithm that turns one into the other. Learning algorithms—also known as learners—are algorithms that make other algorithms. With machine learning, computer write their own programs so we don't have to. (2015, 6)

We can think of the difference between traditional algorithms and machine learners and the algorithms they create with the use of an analogy. Imagine that I wish to build a switchboard system for communication. I want to be able to pick up a phone, say a name and an address, and be connected to that person. To design the system, I build a switchboard that has all fifty states on it, for each state I build a switchboard that lists every city, and then I write a book with the name of every person in that city and the person's particular address. I then hire operators and give them specific instructions. To the operator of the countrywide switchboard I say, "When I call, listen for the name and address. When I say the name of the state, connect to the relevant state and repeat the information I gave you." To the operator of the state-specific switchboard I say, "When the countrywide operator connects to you, listen to what they say, and when they say the name of the city, connect them to the relevant city and repeat the information to the citywide operator." And so on and so forth. This approach to building the switchboard is the traditional algorithmic approach. Alternatively, using a machine learning approach, I tell the learner that it will be fed a name and address and it is to connect me to the person with that name and address. I do not tell it how to go about doing this. It generates a switchboard system of some sort that does the job.

Machine learners are being used to solve problems that traditional algorithms struggle with. Domingos explains that while we understand how to drive cars, much of this knowledge is subconscious and we aren't able to write a traditional algorithm that allows a car to drive itself (Domingos 2015, 6). However, programmers can design a learning algorithm, provide it with desired outcomes and lots of data, and let it design an algorithm for driving. Another example is the use of machine learners in image recognition systems. If I open the photos app on my smartphone, I can search my photos for specific people or objects. For example, if I want to see every picture I've taken of dogs, I might search for "dogs." My phone will, fairly reliably, show me photos of all and only dogs. The algorithm that makes this possible is a machine learner that has been trained on a huge data set using what is called *supervised learning*. The learner is given a bunch of images that are labeled. It then creates an algorithm that attempts to label additional, unclassified images (or images where the learner is not given the classification). This process continues. The learner then develops a new algorithm or model that it uses to identify images as dogs or not. Because the data set that the machine learner has to learn on is very large—think of all the images that Google, the developer of my photo app, has to use as training data—it develops an algorithm that can reliably detect dogs. It can identify which pixels in an image are part of what object and which pixel groups are likely to be classified as dogs, etc. The algorithm that the learner generates for classification takes advantage of rules humans aren't able to program themselves.

Classification is just one of the tricks up machine learners' sleeves, and these tricks can be used to solve a huge variety of problems and accomplish a huge range of tasks. Machine learning is at the heart of many recent technological developments. For example, IBM's Watson not only trounced human competitors in Jeopardy! but is also being employed in a number of other contexts, including medical diagnostics, where it has had success even where human doctors have failed. Or consider Google's AlphaGo, which not only learned how to play Go without being given the rules of the game but is capable of beating the world's very best human players (Sample 2017).

There are many different types of machine learners that differ in how they learn, how they go about creating solutions to problems. One approach to learning or one type of algorithm is an *evolutionary algorithm*. In broad outline, evolutionary algorithms learn and develop algorithms by employing something akin to natural selection. The learner is given a problem and approaches solving this problem by populating a generation of different algorithms to solve the problem. Each member of that generation has a fitness grounded in how well it does in achieving a desired outcome, for example the ability to solve some computational problem. A new generation of programs or algorithms is then populated according to the designed fitness function. So, for example, those programs that were most efficient at solving the computational problem are better represented in the next generation, where they undergo random changes and again try to solve the problem.

It is easy to see how machine learning could, if it doesn't already, generate systems that are dynamic, goal-directed, self-maintaining systems. One concern about artificial intelligence that has received a fair amount of attention from the AI safety community concerns the creation of nonsentient artificial superintelligences that are so dynamic, goal-directed, and self-maintaining that they constitute an existential risk. Bostrom (2016) uses the example of a paper clip-creating artificial intelligence. Imagine a very skilled machine learner, one more powerful than any we currently have, that has been set to one problem: creating paper clips. It is tasked with designing an algorithm that produces the maximum number of paper clips over time. This learner might predict that it needs to safeguard its own power supply, to generate defense mechanisms to ensure that it isn't shut down, or even learn that the most efficient source of some key material is derived from some resource that humans are heavily dependent on. These imagined superintelligent AIs are not sentient or conscious; they are just so good at means-end reasoning or so good at generating algorithms that achieve their ends that they are a threat to humans.

Depending on the problems they are set to, these superlearners might grow or reproduce. Just imagine that a computer scientist of the future wishes to go into the fast-food business but knows nothing of cooking, of building kitchens, of applying for permits, or of when or how to start a new branch. Instead, he creates a machine learner or a stack of coordinated learners to solve the problem. The resulting system doesn't simply output advice, but writes algorithms for designing the spaces to be used, running the construction machines that autonomously build those spaces, for what to put on the menu, for cooking the items on the menu, for taking orders, etc. The learner writes an algorithm that predicts when to start a new franchise and makes a copy of itself to build and run the new franchise.

Finally, it is possible that these dynamic, self-maintaining systems that grow and reproduce might be the result of a process that has all the hallmarks of evolution. There are potentially many dynamic, self-maintaining systems that start out minimally dynamic and barely self-maintaining and, because of the selection pressures imposed on them, offspring of these programs that better achieve goals of reproduction, growth, etc. are the ones that survive to populate the next generation of algorithms. In the end, we might have a system that has all the properties discussed so far in virtue of being selected in the same way that algorithms are. Of course, we might say that this is an artificial selection process, but when we identify this process as "artificial," that's just shorthand for saying that we initiated it and not for any differences in the structure of selection.

What is there to say about the welfare of the algorithms and systems that come about via machine learning, systems that have the historical and ahistorical properties that are often pointed to in living organisms as distinguishing them from artifacts? One option is to countenance the welfare of these artifacts just as we might countenance the welfare of arganisms but reject that typical artifacts meet the relevant conditions for having a welfare to a sufficient degree.

I think this is not a very good option. When biocentrists distinguish artifacts from organisms, they intend to include all organisms, no matter their degree of dynamism, self-maintenance, degree to which they have been subject to selection, etc. There seems a clear inconsistency in demanding that artifacts have to meet some very high standard of these properties in order to count as having a welfare. The biocentrist seems to accept a very low threshold in setting standards for organism welfare, and there is no clear reason why artifacts should have to meet a much higher threshold.

More plausibly, once we see that machine learners themselves (or the systems they generate) might meet all the proposed conditions we tack on to The Basic Account in order to preserve our intuition that nonsentient organisms have a good of their own while artifacts do not, we should just give up on that intuition and stop trying to distinguish them. Nonsentient organisms have a welfare, and this is best understood as being grounded in their being teleologically organized. The reason for thinking that the basic account must be modified is to preserve the distinction between artifacts and organisms in terms of their having or not having a welfare. But the lesson from machine learning, like that of synthetic biology, is that there are artifacts that make it very difficult to preserve the distinction. Coupled with the arguments above that most attempts to distinguish artifacts from organisms fail to do so or do so only by being unacceptably ad hoc, we should prefer the generalized account to the basic account.

Is It a *Reductio*?

What is the appropriate response to recognizing that there is no plausible way to defend the basic account over the generalized account and, thereby, that the most plausible version of the etiological account of teleological welfare entails that in addition to nonsentient organisms and biological collectives, nonsentient artifacts also have a welfare? As discussed at the start of this chapter, many have and will take this as a *reductio* of the etiological account of teleological welfare, thinking that we must find some alternative.

I have already argued that there is no plausible alternative to an etiological account of teleology and, now, that the generalized account is to be preferred to the basic account. If these arguments are sound, then biocentrism is false whether we accept that artifacts have a welfare or take this as a *reductio* of the generalized account. But I don't think that we should take the implication that artifacts have a welfare as a *reductio* of the generalized account.

Notice that a view's having the implication that artifacts have a welfare is not a *reductio* in the technical, logical, sense that it has an implication that is contradictory or metaphysically impossible. If the generalized account implied that there were square circles, that would constitute a *reductio* in this sense. Instead, to the extent that artifacts having a welfare constitutes a *reductio* of the view that implies this, it just means that the conclusion is extremely counterintuitive or so implausible as to be unacceptable.

Are claims that an artifact has a welfare so implausible that any view that entails such a claim must be rejected? One argument in the affirmative might come from how we think and talk about artifacts: it is just part of the way we conceive of them that makes their having a welfare inconceivable or inadmissible. But, as already mentioned, we talk about artifacts in ways that mirror welfare ascriptions in organisms. We talk about what's good for knives, cars, and computers. We might ultimately think that such talk is not literal, that welfare ascriptions to artifacts are ultimately derivative, but apparently ascribing welfare to them isn't so absurd that we scoff at talking as if they have a welfare. Contrast this with attributing complex mental states to artifacts. If I were to claim that a dull knife were sad because of its dullness, this would, I hypothesize, be met with much more confusion; it is more obviously absurd. This suggests that it is not from our intuitions about language use that claims of artifact welfare are beyond the pale.

One argument that the idea that artifacts have a welfare is absurd in the relevant sense is that if we accept that artifacts have a welfare, we would have to countenance artifacts in our moral deliberations. On this view, the generalized account is to be rejected because it implies not only that artifacts have a welfare but that artifacts are morally considerable. This presupposes that it is absurd to think conjointly that artifacts might have a welfare *and* that anything that has a welfare is morally considerable. With respect to the connection between welfare and moral considerability, I argue in the next chapter that we should not think that everything that has a welfare is morally considerable. With respect to the claim that it is absurd to think artifacts might be morally considerable, this view faces the same challenge as the claim that it is absurd to think artifacts have a welfare: it is not absurd in the sense of being contradictory, so we need some other account of what makes it absurd or inconceivable. In my view, this requires the development of cases and corresponding arguments. Not all our judgments about plausibility or intuitiveness are equally good. I submit that it is very difficult, in advance, to judge that artifacts in general can't have a welfare, that this proposition is implausible just upon consideration.

Ultimately, our views about theories of welfare, their implications, and the acceptability of those implications will require balancing. In my view, it is at least as obvious that nonsentient organisms have a welfare as that artifacts don't. Since I've provided an account of welfare that makes good on these intuitions about nonsentient organisms, an account that seems to not only accommodate nonsentient welfare but explain it, an account that meets the subjectivist challenge, I take it that those intuitions are, at least conditionally, vindicated. The fact that artifacts then turn out to have a welfare doesn't dissuade me, certainly not simply because I find that implication icky. There are consequences of artifact welfare that I would find unacceptable, but, I argue, they can be avoided. I revisit some of these methodological commitments and the argument for artifact welfare in chapter 6, where I articulate what I think biocentrists ought to accept in the face of the arguments so far.

6 THE MORAL IRRELEVANCE OF BIOLOGICAL WELFARE

If the arguments up to now are sound, we have reason to accept that nonsentient organisms have a welfare grounded in their being teleologically organized. We also have reason to accept that some artifacts, including corkscrews, can openers, computers, and software programs, have a welfare, and further that some, perhaps many, biotic communities such as eusocial insects and organisms in tight symbiotic relationships have a welfare. In this chapter, I hope to show two further things. First, I will defend a symmetry thesis: if nonsentient organisms are morally considerable in virtue of having interests, teleo-interests, grounded in their teleology, so too are artifacts and communities that have teleo-interests. Second, I will argue that nonsentient organisms are not morally considerable. This will complete the argument against biocentrism. By showing that biocentrism is an unstable position, that one committed to it has not extended the sphere of moral considerability far enough, and that further expansion is untenable, I will have shown that biocentrism must be abandoned.

A No Morally Relevant Differences Argument for Teleocentrism

Teleocentrism is the view that all entities with teleo-interests, interests or a welfare grounded in the etiological account of teleology, are morally considerable. I do not know of any teleocentrists, and, as will be argued, it is a position we ought to reject.[135] But

---

135. There are people comfortable with attributing interests (or at least moral status) to mere machines (Gunkel 2012), robots (Gunkel 2014), and even data (Floridi 2002), but it isn't clear to me whether they include all artifacts that are teleologically organized.

does the biocentrist have any basis for rejecting teleocentrism, any grounds on which to restrict moral considerability to nonsentient organism despite artifacts' and biotic communities' also having a welfare grounded in interests of the same kind as nonsentient organisms?

The answer is no. The same strategy, the strategy of extension, that biocentrists working in the welfare approach employ to argue for an extension of moral considerability to nonsentient organisms can be employed to expand the sphere of moral considerability even further. Recall the strategy of extension as discussed in chapter 1: an anchoring class and a target class of entities are identified. It is argued that entities in the target class have interests of a kind similar to entities in the anchoring class, and, finally, it is argued that there is no morally relevant difference that would justify recognizing the moral significance of the interests had by those in the anchoring class while rejecting the moral significance of entities with interests of similar kinds in the target class. In order to argue for teleocentrism, let the anchoring class be nonsentient organisms and let all other entities with teleo-interests be the target class. The biocentrist is already committed to the moral significance of the entities in the anchor class, and the previous chapters have argued that artifacts and biotic communities of the relevant sort have a welfare, grounded in teleo-interests, of the same type as the entities in the anchoring class. All that is left to do to show that the biocentrist must convert to teleocentrism is to argue that there is no morally relevant difference between the entities in the anchoring and target classes that would justify differential consideration of the respective interests of those entities.

This task seems straightforward. What differences are there that the biocentrist could cite as morally relevant? Of course, there are differences. It is true, for example, that entities in the anchoring class are alive while entities in the anchoring class, at least most artifacts in any case, are not. It is also true that that anchoring class includes only organisms, while the target class, for the most part, does not. In the case of these differences, we've already seen that they don't make a difference with respect to whether the relevant entities have a welfare; being alive is not a necessary condition for having a welfare, nor is being an organism. According to the strategy of exclusion, the prospects for helping the biocentrist to distinguish the anchoring class from the target class were to be found in these properties' being necessary for having a welfare vis-à-vis their being necessary for being genuinely teleologically organized. The biocentrist owes us an explanation for why whatever difference they cite as morally relevant is so despite not being relevant to which things have a welfare.

There is a reason that sentientists and biocentrists adopt the strategy of exclusion as opposed to trying to define the exclusion class in another way. For those committed to the correlate of welfarism, this alternative strategy just isn't possible. If things in the purported exclusion class are judged to have a welfare, they are morally considerable. Even for those that admit a conceptual distinction between having a welfare and moral considerability, like Taylor, they argue that having a welfare is sufficient for moral considerability. The reason for this is simple: Once we judge welfare of a particular type is morally considerable, what grounds could we have for rejecting the moral considerability of the genuine interest in other things? As Singer (2009) argues, there are descriptive differences between people of different races and sexes, but none of those provide any reason to discount or dismiss the moral considerability of shared interests, any reason not to treat like interests alike. The same can be said of attempts to distinguish the moral considerability of some entities with teleo-interests, those of organisms, from others. I contend that in showing that the differences between the anchoring class and the target class are not relevant to whether entities in these classes have a welfare, those differences are also shown not to be morally relevant differences. To assert the moral relevance of one of those differences will be arbitrary or question-begging. In a sense, the biocentrist falls prey to the same trap that Singer's argument for sentientism falls prey to. Recall, Singer argues for drawing the line at sentience as nonarbitrary precisely because it draws the line at necessary conditions for having a welfare. Biocentrists deny this and assert that such a conception of interests is too narrow. However, biocentrists have made use of a similar strategy to deny the moral significance of the "interests" of artifacts and communities by denying that such entities have interests at all. Now that we see that they do, it looks as if biocentrists commit something akin to speciesism; they arbitrarily favor the interests of a narrower class of entities than is rationally required by their commitments.

This leaves the biocentrist in the same precarious place as those who wish to defend anthropocentrism in the face of these sorts of arguments; it is open for them to identify a difference between the classes and defend that difference as morally relevant, but that is a significant burden. The biocentrist cannot merely assert that some identified differences are morally relevant. That would be to beg the question. Furthermore, the relevant difference cannot be one tied to our attitudes or desires regarding entities in the different classes. First, not everyone values, cares about, or has some other positive attitude toward nonsentient organisms but not toward entities in the target class. Second, if the moral status of the anchoring class is contingent on our caring about the entities in that class, then their moral status is not direct; the welfare of nonsentient organisms might make certain attitudes toward them coherent, such as an attitude to promote or protect their welfare, but it won't be the basis of their moral considerability.

Absent some compelling argument for asymmetry, there is a presumption in favor of symmetry. We have no reason to distinguish between the kinds of things—nonsentient organisms, artifacts, and communities—in terms of the moral significance of their welfare. In other words, if nonsentient organisms are morally considerable, so too are the relevant artifacts and biotic communities.

The Case against Teleocentrism

In response, the biocentrist may acquiesce—accept the expansion of moral considerability to all entities with a welfare grounded in the etiological account of teleology. In fact, some holists, prior to seeing exactly which biotic communities turn out to have a welfare, almost certainly would have been comfortable with the consequence of expanding biocentrism to include nonsentient collectives (though they probably won't like the expansion to artifacts). Perhaps some holists will feel that an "I told you so" is in order, even if teleocentrism doesn't yield the conclusion that traditional wholes are morally considerable. In any case, teleocentrism should be rejected. In what follows, I provide an argument against teleocentrism for those biocentrists who are unwilling to give up their commitment to the moral considerability of nonsentient organisms and so might be tempted to bite the bullet of teleocentrism.

As in other chapters, I want to be fairly generous to the biocentrist. While it might seem that teleocentrism is just obviously false, that the very fact that a biocentrist would be forced to adopt the view is reason to reject the commitments that ground biocentrism, I do not think that is the best argument. Instead, I think it is best to draw out specific consequences of teleocentrism that I hope biocentrists will reject and in doing so to force them to reject their commitment to the moral considerability of all those things that are teleologically organized. Thus, the argument against teleocentrism will, like many arguments in ethics, draw on cases—thought experiments meant to isolate variables of interest and elicit a clear judgment about the case that can then be used to make determinations about some hypothesis, principle, or claim. I do not intend here to defend the methodology of reflective equilibrium or the use of thought experiments in philosophical reasoning. Suffice it to say that I think that ultimately intuitions about cases, principles, and the

like are ineliminable from the doing of philosophy, including doing justifica-
tory work, and thereby, from engaging in most forms of rational inquiry. In
the words of Kripke:

> Of course, some philosophers think that something's having intuitive
> content is very inconclusive evidence in favor of it. I think it is very
> heavy evidence in favor of anything, myself. I really don't know, in a
> way, what more conclusive evidence one can have about anything, ulti-
> mately speaking. (Kripke 1982, 42)

However, in Thomson's words: "There are cases and cases" (1971, 58).
While many might happily accept that intuitions play an important role in
theorizing, the cases employed here will be admittedly strange. They involve
highly unrealistic scenarios and depend on judgments about such scenarios.
One might rightly question whether such cases produce reliable judgments,
whether they may be the basis of legitimate ethical theorizing. In my experience, many students are fine with some cases, but balk at others such as
Thomson's famous case involving people seeds that, if they embed in one's
carpet, develop into people (Thomson 1971, 59). I expect many ethicists, especially practically minded environmental ethicists, might feel similarly. In
light of this, before turning to the cases used to argue against teleocentrism,
it seems prudent to canvas some alternative approaches to arguing against
teleocentrism. To the extent that these other approaches are successful, that
is all to the good. For my part, biocentrists may reject their view for any of the
provided reasons. My hope is that by canvasing all of these, I provide reasons
for every biocentrist to abandon ship.

Self-Evidence

Some claims or principles seem to be self-evident; our understanding them
commands our assent. Descartes thought of his cogito as self-evident in this
way, and many of us believe that the law of noncontradiction is self-evident.
Many have believed that certain other principles or axioms of classical logic,
such as the law of excluded middle, are self-evident in this way. Within moral
philosophy, many have seen consequentialism or some component of it is as
self-evident: How could something that produces the *best* outcome be wrong?
Feldman's axiological commitment to hedonism, discussed in chapter 2,
seems to be based on what he takes to be a self-evident fact about welfare.
While I do not intend to defend a theory of self-evidence—a theory about
whether all self-evident truths are, for example, analytic or how to understand

our a priori faculties such that we recognize such truths—I don't deny that some claims might be justified in this fashion.[136]

The question is whether the truth or falsity of teleocentrism is self-evident. It's truth certainly is not. While some biocentrists might think it self-evident that certain claims used in arguments for biocentrism are self-evident, maybe including the claim that plants have a good of their own, every biocentrist recognizes that an argument is required to defend the view. The fact that many philosophers have seen fit to write articles and books in defense of bio-centrism belies the claim that it, let alone teleocentrism, is self-evidently true.

Is teleocentrism self-evidently false? If you had asked me a decade ago, be-fore I was steeped in the literature on teleology and etiological theories of func-tion, whether I thought teleocentrism was false. I believe I would have said, upon only a little reflection, that it was. So there is a sense in which my past self took the view to be false upon reflection on the view. On the other hand, I don't think this shows that there is reason to think the claim is self-evidently false. I might have had a hunch that there were good reasons to dismiss the view, or I might have thought I could come up with some counterexamples to show that the implications of the view were implausible. In fact, the falsity of teleocentrism seems nothing like the self-evident falsity of the denial of the principle of noncontradiction.

Even if teleocentrism is not self-evidently false in the sense that it is con-tradictory or such that the view can immediately just be seen to be false, it might be self-evidently false in a weaker sense, that mere reflection on the view provides a reason to reject it; i.e., it is strongly counterintuitive. Since teleocentrism, as defined, includes a commitment to simple artifacts both having a welfare and being morally considerable, I agree that it is counter-intuitive, but this is not automatically a sufficiently strong reason, at least for those previously committed to biocentrism as defended within the welfare approach, to reject the view. After all, such individuals, I've argued, are really committed to teleocentrism on the basis of views which they take or should take to be plausible. Of course, perhaps the consequence of teleocentrism is too much for some biocentrists to bear. The main argument of this book is that we must give up biocentrism, and if this argument succeeds without the need to appeal to the cases I describe below, then the argument is perhaps stronger for it.

---

136. For some recent discussions of the a priori and intuitions see Bengson 2010, 2013; Cappelen 2013; Chalmers 2014.

Appeal to Principles

Another perfectly legitimate way to evaluate a philosophical claim is to argue for or against it by appealing to other principles that bear on that claim. Obviously, the strongest form of this sort of argument for (or against) a claim is to derive the claim (or its negation) from some principles we take to be self-evident, or at least very hard to deny, but philosophers also often consider just how well a given claim fits against a background of other principles and claims that they accept. This is common in, for example, arguments from the best explanation. So we might not need strange cases in order to decide the truth or falsity of teleocentrism insofar as it is inconsistent with or doesn't fit well with other principles that we accept.

But which principles will do the work here? The most obvious principles that might seem to do the trick have already been discussed. One might try to argue that teleocentrism is inconsistent with some normative theory that we take to be true. If teleocentrism is inconsistent with hedonistic utilitarianism (it is), then we have reason to reject teleocentrism. But, as was argued in chapter 2, normative theories do not settle questions of the bearers of moral status. A normative theory in the structural sense implies nothing about the question of who or what has moral status, and questions about the bearers of moral status are settled independently of and, in some ways, prior to questions about normative theories in the specific sense. As it turns out, if teleocentrism is true, we know that hedonistic utilitarianism is false and not the other way around.

I'm not sure what other principles or nexus of principles and claims might be used to show that teleocentrism is false independently of appealing to a set of cases that develop the consequences of teleocentrism in ways that provide grounds for rejecting the view. But I do know that there is a nexus of principles and claims, absent such cases, that suggest that teleocentrism is true; it is whatever nexus of principles and claims motivate biocentrism in conjunction with the principles and claims defended in previous chapters to arrive at teleocentrism!

Simple Cases

Even if we have to appeal to cases or thought experiments to undermine teleocentrism by exposing implausible consequences of the view, it doesn't follow immediately that we must consider cases that are highly idealized, those that demand we consider very unrealistic states of affairs. Unfortunately, the prospects for identifying simple cases that will do the work are poor. Why is this? Consider the following judgments:

1. I do nothing morally wrong in letting my car rust.
2. The death of a houseplant is worse than the breaking of the pot it is planted in.
3. Playing video games that result in the destruction of digital avatars is permissible.

There are many cases very close to real life where these judgments about those cases would be plausible. Imagine that I'm extremely busy in the pursuit of worthwhile aims or that I am extremely poor, and one of these explains why I let my car rust. In such a case, it does seem that I do not, at least not all things considered, do anything morally wrong in letting my car rust. Imagine that a houseplant is a rare and beautiful flower but the pot that it is planted in is common and easily replaced. Imagine that the video game being played is one that is used for therapeutic purposes such as helping trauma patients with PTSD reprocess memories. Cases that yield these judgments can be used to show that teleocentrism is false only if these judgments are inconsistent with teleocentrism. But, of course, they are not. If teleocentrism is true, then the interests of my car, the pot, and the digital avatars are morally significant, but it doesn't follow that such significance provides a reason to think we act wrongly by letting the car rust, the pot break, and the avatar be destroyed.

Furthermore, there are simple cases where these judgments appear to be false. Imagine that I am rich, I have multiple cars, and there is a very needy individual to whom I have an obligation that would be met by my providing the person a well-maintained car. In such a case, it would seem, I have an obligation to keep my car in good shape or not to have let it fall into disrepair. Again though, this judgment tells us nothing about the truth of teleocentrism, at least not unambiguously, since this obligation can be explained as an indirect obligation to maintain the car on the basis of a direct duty I have to someone else. If teleocentrism is true, I have a pro tanto reason to maintain the car in virtue of the car's interests, but that reason might be screened off by my direct obligations to others. One of the reasons that Varner's arguments for biocentrism are uncompelling is that the case used to establish that, for example, his cat Nanci has biological interests is too much a real-world case. Yes there is a plausible judgment made that going outside might promote or frustrate Nanci's interests. But the case isn't sufficiently described so as to rule out alternative explanations of this judgment that don't appeal to Nanci's biological interests.

If we want to determine whether entities are morally considerable solely on the basis of having teleo-interests, i.e., if we want to determine whether such interests impose on us a pro tanto obligation to promote or protect those interests, we must develop cases where the only relevant variables determining our obligations are those interests.[137] We need cases where our judgment that we have an obligation toward nonsentient entities cannot be explained in terms of some other obligations. But these cases will necessarily be contrived. Just as a telling scientific experiment most often requires we create an environment that removes most of the complexity of the world, so, too, does a good thought experiment. The world is a complex nexus of interests and obligations; to test which interests give rise to obligations, we must imagine different and sometimes distant worlds.

The Case(s) against Teleocentrism

Consider how a thought experiment might figure into an argument against teleocentrism via the following schema:

1. If teleocentrism is true, we have an obligation to Φ in C (that we would not otherwise have).
2. We do not have an obligation to Φ in C.
3. Therefore, teleocentrism is false.[138]

If an argument of this form is to be cogent, a thought experiment must do two things. First, it must describe a circumstance, C, in which the welfare of

---

137. Readers will recall that in chapter 2, questions of the bearers of moral status were distinguished from questions of how bearers of moral status are to be accommodated. Here it seems that I'm running those questions together and that these questions are related. A thing is a bearer of moral status if and only if it is to be accommodated in our moral deliberations (for its own sake). Thus, if we can show that a thing does not require any accommodation, i.e., if we can never have any moral obligations to it, we know that it is not a bearer of moral status. The reason for distinguishing these questions in earlier chapters is that to precisely explain how a bearer is to be accommodated requires the help of a normative theory. Here, however, the argument will hinge on the claim that we have no obligations whatsoever, no matter which normative theory is true, toward entities with teleo-interests. So, again, we can proceed without settling questions of normative theory.

138. The decisiveness of any particular instance of this schema will depend on how justified we are in thinking that it is sound, which will be a function, for those who employ the method of reflective equilibrium, of how plausible our judgments about the cases that support the premises tempered against any principles one takes to be justified but have implications inconsistent with those judgments.

an entity grounded in its teleo-interests *grounds* an obligation for agents to act, Φ, a certain way. To ensure that it is the teleo-interests of some entity that are in fact grounding, i.e., explaining or making it the case that we have the obligation, C should be such that we would not otherwise have an obligation to Φ. In other words, a good thought experiment describes a circumstance with determinable, unambiguous consequences of teleocentrism. Second, the case must elicit the judgment that agents do not have the obligation entailed by teleocentrism in the circumstance.

Moral philosophy has furnished us with a variety of thought experiments to use as models for developing the challenge to teleocentrism. In what follows, I modify several of these thought experiments for the purposes of filling out the above schema in a way that undermines teleocentrism.[139]

The Last Woman

To start, let us consider a parallel to the most famous case used to defend biocentrism, the Last Man. In these cases, we are to imagine a nearly empty Earth whose sole inhabitants are a tree and a man. We must assume that the tree has whatever is necessary to continue to live once the man has died and that any interest the man has in cutting down the tree is trivial (i.e., he doesn't need the wood for fire to survive, etc.). We can imagine that the last man is dying or will very soon die but that the tree will live out the rest of its life on the barren planet. We are to consider whether the man acts wrongly if he were to cut the tree down, i.e., whether he has an obligation to preserve it. Many environmental ethicists think the answer is yes and this judgment is sometimes a cornerstone in arguments for biocentrism.

Even if Last Man thought experiments elicit the judgment that we have an obligation to leave the tree alone, it isn't clear that the case is one in which that obligation is grounded in the moral considerability of the tree. The judgment might be explained in aesthetic terms; perhaps we just find a world with a tree on it more beautiful than a world completely devoid of life. If we vary the last nonsentient organism, imagine, say, a mosquito or a tick, perhaps our judgment changes, thus showing that it is not the moral considerability of the tree

---

139. Each of the cases involves artifacts rather than biological collectives. This is for a reason already mentioned; it is hard to construct cases where a collective is harmed without also harming individual organisms (at least not without some heavy presuppositions about what exactly is owed to individual organisms and an argument that those duties aren't violated by, for example, removing them from their natural habitat). Since harms to individuals and to collectives are so tightly correlated, it is hard to disentangle the various judgments we might have about cases involving both types of entity.

that grounds the last man's obligation to leave it be. Or perhaps the judgment is explained by some judgment about the character of the last man, that he wastes what little time he has left by engaging the tree.

In contrast to Last Man, consider a series of cases involving the last woman. The last woman lives, for all intents and purposes, on a completely barren planet. If we can't even make sense of a world where a human can live independently of organisms and environments that make the planet habitable, we can imagine that the last woman has no chance to affect or interact with those living systems, her portion of the planet is barren, and her powers to alter things outside that portion of the planet is nil. However, there are artifacts of varying kinds on her barren stretch of the planet. We can think about how the last woman might interact with these artifacts and whether and what her obligations might be.

Variant 1: Shutting Down
The last woman has built a home for herself on a barren landscape. She has many artifacts that help her to live as comfortable a life as she can on a barren planet. The last woman knows her life is coming to an end. In her last days, she removes the battery that was powering her home and turns off the water pump that was providing water to the home. She then shuts down her other life support systems and passes away.

Variant 2: Hubris
The last woman wanders a barren patch of her planet. She knows that she will soon die because the life support systems of the planet are breaking down. In her final days, she encounters various artifacts: a can opener, a mousetrap, and a laptop. Each artifact reminds her that it was the hubris of humanity that lead to the barrenness of her planet, that it was an overreliance on technology that was their ultimate demise. In anger, she smashes each of these artifacts.

Variant 3: Virus
The last woman has built a home for herself on a barren landscape. She has many artifacts that help her to live as comfortable a life as she can on a barren planet. Among these artifacts is a computer that is powered by solar panels and a battery. She discovers among the programs on the computer a piece of software that seems to have been programmed to create copies of itself and spread them to other computers. The developer left a note inside the program that the goal of the program was to make itself the most abundant program on the planet. Whenever the computer is on, it makes copies of itself. The woman deletes these

copies whenever it would slow down the machine enough to under-mine her enjoyment of the computer. Eventually, she no longer gains any enjoyment from the computer. She turns off the computer forever.

Variant 4: Unwinding

The last woman wanders a barren patch of her planet. She knows that she will soon die because the life support systems of the planet are breaking down. In her final days, she encounters various artifacts: a can opener, a mousetrap, and a laptop. She collects each artifact. In the evenings, while recalling fond memories, she mindlessly takes apart each artifact, rendering each useless. She is not angry when she does this; instead it merely serves to satisfy a small desire to be occupied. When she runs out of artifacts to destroy, she twiddles her thumbs and her mind remains as at ease as when she was deconstructing artifacts.

Consider the following claims about potential obligations the last woman might have in each of these variants:

1. The last woman has an obligation not to shut down her support systems in Shutting Down.
2. The last woman has an obligation not to destroy the found artifacts in Hubris.
3. The last woman has an obligation to leave the computer running in Virus.
4. The last woman has an obligation to avoid breaking apart the artifacts in Unwinding.

By my lights, all of these claims are false; the last woman has no such obligations to the artifacts described in these cases. I arrive at these judgments simply by careful reflection on the cases themselves. If a reader does not share these judgments, then in order to convince such a reader, I must employ different cases.

Even if one agrees that the last woman does not have the ascribed obligations, it remains to show that teleocentrism entails otherwise; failing that, these cases are just irrelevant to the truth of teleocentrism. Teleocentrism will unambiguously and determinately imply that we have obligations in these cases just in case (i) the last woman's actions contravene the welfare of the artifacts in the case, and (ii) the last woman has no sufficiently good reason to contravene those interests, a reason that overrides the interests of the relevant artifacts.

In Shutting Down, it isn't obvious that the last woman contravenes the welfare of the artifacts in her home. This depends on an account of precisely how we analyze artifact welfare. For example, at one level of description, the ends of a water pump are to pump water and so shutting it down undermines its ability to achieve its ends. If this is the level of description relevant to the analysis artifact welfare, then the last woman seemingly does contravene the welfare of the water pump. However, if the relevant level of description involves specific, abstract aims of the designer, this might not be the case. If, for example, the last woman designed the water pump herself with the intention that it serve her until her last days, then it seems as if the water pump has achieved its ends, it has no more ends or purpose, and so shutting it down does not contravene its welfare. In Hubris, it isn't clear that the woman doesn't have a sufficiently good reason to contravene the welfare of the artifacts in the case (assuming she does so). After all, the last woman's anger might very well be justified and the destruction of the symbols of hubris might be an appropriate expression of that anger.

Virus and Unwinding are built so as to avoid these sorts of challenges. In Virus, the ends of the designer and of the computer program are specified in such a way that it is supposed to be clear that the last woman contravenes the welfare of the computer program, and once she no longer gets joy out of the computer, it seems she no longer has a sufficiently good reason to contravene those interests (if they are morally considerable). Similarly, in Unwinding we can imagine that some of the artifacts that she destroys to pass the time have a welfare that is contravened and furthermore that whatever reason she has for destroying those artifacts is trivial.

One way to respond to these examples is to claim that even the most trivial interests of humans trump the interests of artifacts such that any reason at all is sufficiently strong to justify contravening the interests of artifacts. However, this saves teleocentrism as a view about the bearers of moral status at the cost of undermining it as a view with any normative implications. Call this view toothless teleocentrism. It is toothless because every intentional action, plausibly, is grounded in some reason, and so, one never wrongs an artifact by contravening its welfare while acting intentionally. On such a view, it isn't clear what it would mean to attribute moral considerability to artifacts (or nonsentient organisms) since for all intents and purposes the welfare of such beings never has to figure into our moral deliberations; whatever "moral considerability" means on this view, it is a moral considerability without teeth, a sense of moral considerability that no biocentrist can accept since biocentrists think their view carries implications for our actions regarding, at least, the environment.

Biocentrism has been criticized by others on grounds that it provides no ethical guidance, that the sheer number of things that have interests and what would be required to take them into account makes the view untenable. My objection to toothless teleocentrism differs from these criticisms, and it is worth discussing how.

One criticism in the general vicinity is that once everything satisfies some definition for being morally considerable, that definition becomes untenable. The thought is that if everything is morally considerable, the notion of moral considerability doesn't identify anything special about those that have it.

There are, of course, useless distinctions. It would be pretty useless to try to mark a distinction between rulers that have a length and rulers that have no length. But attempts to offer an account of moral status, to offer an analysis of what makes something morally considerable, etc., aren't, at least not primarily, attempts to draw a distinction between those things that are morally considerable and those that are not. Yes, an analysis of, say, moral considerability would allow this, but the point of such an analysis is to tell us which things ought matter in our moral deliberations, and there is no reason to think "which things" couldn't be "all the things." Just imagine that there is an afterlife. In it, the only things that exist are souls. Souls can interact, they have emotional states, they can be made happy or sad, enjoy conversations, and suffer from taunts. I take it that each soul is morally considerable and find nothing strange that everything that exists is morally considerable.

A second criticism of biocentrism along these lines is that if biocentrism is true, there are just too many interests to care about, no action is possible that does not contravene innumerable interests, and so decision-making becomes impossible. This criticism has been taken up by several biocentrists such as Varner (1998, chap. 4) and Attfield (Attfield 1995). Responses to this criticism attempt to undermine the objection by offering accounts of different types or strengths of interests, and by laying out principles or methods for trading off interests against one another. Some of these solutions will be better than others, but there is no in-principle problem with solutions of this type, no general reason to believe that this is a problem for biocentrism. Ranking interests and developing procedures for dealing with conflicts is, after all, something sentientists must grapple with as well (see VanDeVeer 1979), and the fact that we take something to be morally considerable can simply serve as a constraint on which normative theory and decision-theoretic tools we use to derive the implications of that moral considerability.

These criticisms trade on the idea that there just is no way to be responsive to the interest of nonsentient organisms, that trying to do so would make acting morally impossible. The argument against toothless teleocentrism is not that teleocentrism is too burdensome, but that it is not burdensome at all. The judgments about cases discussed above are meant to show that we can permissibly trounce the teleo-interests of artifacts, but these interests are of the same kind (and to the extent that we can talk about the strength of these interest, the same strength) as the teleo-interests of nonsentient organisms. This is not a case where we can prioritize the interests of nonsentient organisms because they are of a different kind or where we can ground their greater strength in the way we can ground comparisons of strength in the case of interests grounded in desires of different importance or strength. So it seems to follow that we can trounce all teleo-interests—that the "moral considerability" of toothless teleocentrism is just a title.

Sterba (1998) has offered an account of what distinguishes artifacts from organisms that might be adapted to avoid this objection. While Sterba thinks that most current artifacts lack a good of their own because of their derivative nature, he doesn't think this is necessarily so, that no artifact could be so organized so as to count as having a good of its own. Yet he still thinks artifacts are not morally considerable. The difference between organisms and artifacts is grounded in what he calls "the prerogatives of and constraints on moral agents in their relationship with other living things" (Sterba 1998, 372). For Sterba, to recognize that an organism is morally considerable is not just to recognize that we have to take it into account in some way—it is not just a judgment that a thing is a bearer of moral status; instead he claims that "when moral agents recognize beings as having intrinsic value, they simply recognize that these constraints apply to their interactions with them" (Sterba 1998, 373). The specific constraints that he thinks apply to other living things is a contextual constraint not to interfere; we have an obligation, under some sets of circumstances, to let other living things be. This constraint makes no sense with respect to artifacts. To the extent that they have a good, it so substantially overlaps with our own that there can be no requirement to leave them alone. To do so would not be good for them (Sterba 1998, 374).

The upshot of Sterba's argument for the current discussion is that while artifact welfare really is normatively toothless, it doesn't follow that teleocentrism is toothless; the welfare of nonsentient organisms (and perhaps collectives) can carry normative weight even while that of artifacts does not. This seems to allow biocentrists to have their cake and eat it too; they can accept that their view really must be teleocentric and yet deny that we have to be responsive to the good of artifacts.

Ultimately, this solution fails. First, given the etiological account of teleological welfare, there will be many cases where we can contravene the interests of artifacts, where our interests don't substantially overlap with their ends. Several of the cases above are plausibly instances of this kind. But, if that is so, then so long as the pro tanto requirement to leave living things alone is grounded in their interests rather than in some indirect duties, that same requirement plausibly applies to artifacts. Second, this argument depends on our seeing questions of the bearers of moral status being settled by questions of normative theory. I've already argued for epistemic independence, but this argument concerns a kind of metaphysical or explanatory dependence. We can understand the argument as an argument that a normative theory that generates plausible constraints on human actions will not constrain those actions solely on the basis of facts about artifact welfare. But to narrow our choice of normative theories in this way begs the question; it assumes that there is some morally relevant difference between artifacts and organisms.

The Teleo Trolley

Essentially, a Trolley case is one in which we must decide how to respond to a runaway trolley, whether to allow it to continue on its course down a track resulting in some bad consequence or to divert its course via some means which will result in some other bad consequence.[140] Traditionally, Trolley cases have been used to elicit judgments that tell for and against different specific normative theories (and the moral relevance of various distinctions such as killing vs. letting die). For example, the most famous examples ask whether we are morally obliged to divert a train from a track that will kill five innocent people to a track where only one such person will be killed, or whether we are obliged to stop a runaway trolley by pushing an innocent person in front of the trolley in order to save five innocent people further down the track. Whereas in the first instance the judgment that it is obligatory to divert the trolley is often taken to support some version of utilitarianism, the second case is often taken to be a counterexample to such views since it elicits the judgment that we act wrongly by pushing someone in front of the tracks.

Since the mechanism by which Trolley cases do philosophical work is via eliciting judgments about obligations, we can modify these cases in a way that helps us fill out the schema above, yielding an objection to teleocentrism. There are two different types of Trolley case that I will used to argue against teleocentrism. In one type of case, call these cases "different interest cases," the different tracks or choices are between a thing (or things) that has (have) only teleo-interests and a thing that has moral considerability grounded in

---

140. For discussions of Trolley cases and examples see Kamm 2015.

some other kind of interest, e.g., a sentient being. Here is an example of such a case:

Startled Dog

There is a runaway trolley heading down a track toward a dog. If the train proceeds down the track, the dog will be startled by the approaching train but will be able to run off the tracks and avoid being hit. It is distant enough that the dog has not yet seen it, and I can indeed divert it before the dog will become aware of the train's existence. If the train is diverted, a simple artifact will be completely destroyed.

A second type of case, "same interest cases," involve a conflict between two different entities which both have only teleo-interests. Consider the following:

Computer vs. Corkscrew

There is a runaway trolley heading down a track toward a working notebook computer. If the train proceeds down the track it will destroy the computer beyond repair. It is possible to divert the trolley down another track, but on this other track lies a perfectly usable corkscrew. If the trolley is diverted it will destroy the corkscrew beyond repair.

These are, I think, both silly cases. The first is silly, by my lights, because my obligation is so obvious. The fact about the artifact's being on the track is relevant only because it tells me that it is the only thing on that track, making my obligation very clear: the trolley should be diverted. The second case is silly, by my lights, because there seems clearly to be no obligation whatsoever, at least not one grounded in the facts about the welfare of the artifacts on the tracks. I suppose that I intuitively think we should preserve the computer over the corkscrew, but this is only because I recognize the value of computers. It is easy to think of variants of these cases that control for these kinds of indirect reasons for favoring one choice over another.

Teleocentrism is problematic precisely because it entails that these cases aren't silly, that there is a genuine conflict that makes these choices or some close variants difficult. Teleocentrism implies that there are some circumstances in which we should allow the dog to be startled solely because of the interests that would be contravened by allowing the destruction of the artifacts were the trolley diverted. If this doesn't seem clear in the case of Startled Dog, I invite you to ratchet down how the interests of the dog that are contravened as much as you like, to make the way that the dog is impacted bad for the dog but only trivially so—perhaps it is annoyed at having to move out of the way of the train—and ratchet up the total welfare that is contravened by diverting the trolley—perhaps a billion calculators are destroyed or perhaps one computer with millions of teleo-interests constituted by the software that it runs. So long as we control for indirect effects that implicate other interests or moral values, I contend that such cases remain silly, and our obligation remains clear: divert the trolley.

The same thing can be said about same-interest cases. The teleocentrist is committed to there being an obligation in such cases. What that obligation is will depend on how many interests are at stake or the relative importance of them, but there are artifacts that we can put on the different tracks that obligate us to choose to divert (or not), not on the basis of whether the artifacts are rare, valuable, important, beautiful, hard to replace, etc., but solely on the basis of the teleo-interests at stake. This seems to me to be obviously false.

The Teleo Monster

The Utility Monster is a famous thought experiment developed by Nozick (1974) as part of an argument against various forms of utilitarianism. The utility monster derives much greater pleasure (enjoyment, etc.) than we do from otherwise similar experiences. The monster is constructed such that any resource we might use to generate positive welfare value for ourselves would generate more of such value for the utility monster. If utilitarianism is true, then no matter how good some resource use would be for us in terms of making us better off, we are obligated to instead use those resources to generate pleasure for the monster.

Utility Monster, like Trolley cases, works by eliciting judgments about obligations, and it can also be modified to develop an argument against teleocentrism. Consider the following:

The Antihuman Teleo Monster
There is a computer with various pieces of software with various ends. One piece of software has the end of writing additional software with ends that are contrary to the ends of humans. This piece of software generates many other pieces of software that have as their end, or as a means to their end, significant welfare diminishment of human beings.

Or, this variant:

The Neutral Teleo Monster
There is a computer with various pieces of software with various ends. One piece of software has the end of generating new software. It

generates a vast number of programs with various ends that are satis-fiable only by their being provided power via the pedaling of a series of bicycles attached to the computer that contains the software programs.

We can ask about either variant: do humans have any obligations toward these computers and the programs they contain? Would it be wrong to power down either teleo monster? To disable the software that generates additional programs? To fail to pedal the bikes? If teleocentrism is true, there is some number of programs in these machines, some number of teleo-interests that would be satisfied, and some number of human interests we could sacrifice to satisfy those teleo-interests that would obligate us to allow our welfare to be significantly diminished (in the case of Antihuman Teleo Monster) or obligate us to dedicate some significant number of hours (in Neutral Teleo Monster) to satisfy these teleo-interests. There is no number of people so small with interests so trivial that even an infinite number of programs would *require* those humans to sacrifice their interests for the sake of the teleo monsters.

This objection (as well as some variants of Teleo Trolley) depends on the assumption that teleo interests are in some sense additive, that somehow the collective welfare of a large number of teleo-interests imposes a stronger ob-ligation than a smaller number. Tom Regan has argued against this kind of additivity. According to his "worse-off principle":

> Special considerations aside, when we must decide to override the rights of the many or the rights of the few who are innocent, and when the harm faced by the few would make them worse-off than any of the many would be if any other option were chosen, then we ought to over-ride the rights of the many. (Regan 1983, 308)

According to this principle, if one must decide to override the rights of one person in such a way that, for example, that individual will be severely injured or to override the rights of the many, even very, very many, such that they will all suffer a paper cut, we ought to opt in favor of the paper cuts. In other words, the interests of the many aren't additive.

Denying that interests are additive in the sense that contravening a larger set of interests does not immediately generate a stronger obligation not to con-travene those interests will not help the teleocentrist avoid the implications I've claimed the view has. If even the most trivial interest of a sentient being, an interest grounded in something other than teleology, is sufficient to jus-tify contravening the most basic interest of an artifact, *and* obligation isn't strengthened by additivity, then Teleo Monster is irrelevant to the case against teleocentrism.[141] But this is only because we don't need a teleo monster, but only a teleo toddler, to show that teleocentrism is false. We have no obligation to sacrifice any resources at all to satisfy the single basic interest of a single piece of software, at least no obligation grounded in the moral considerability of that piece of software. Teleocentrism has again become toothless.

What follows from the above cases and how they fill out the above schema? By my lights, the above cases serve to undermine teleocentrism. Whatever reasons, arguments, etc. there are in support of biocentrism as defended in the welfare approach, they do not survive the overall argument presented so far. The fact that the most plausible version of that view, one that fills out the details of how nonsentient organisms have a welfare in a way that is nonarbitrary, nonderivative, etc., requires that we accept teleocentrism, combined with the fact that teleocentrism yields these extremely implausible implications, is sufficient to reject teleocentrism.

What Does the Argument against Teleocentrism Tell Us?

There are two different conclusions that can be drawn from the case against teleocentrism. This is because teleocentrism, since it is just an expansion of biocentrism, carries two commitments: (i) that genuinely teleologically organized entities have a welfare, and (ii) that welfare (or its bearers) are morally considerable. Given this, one can accept that there is in fact a meaningful sense of welfare grounded in teleology but reject that the sorts of interests that constitute such a welfare are morally significant, *or* one can take the implications of teleocentrism as providing a reason to reflect back on the plausibility of the etiological account of teleological welfare and reject outright that there is any genuine sort of welfare or interest grounded in teleology. Which is the most plausible option?

The answer to this question will ultimately depend on weighing the theoretical costs and benefits of each option against one another. My view is that we should accept the etiological account of teleological welfare but reject the moral considerability of those that only have welfare grounded in that account. Many of the reasons in favor of this option have already been argued for or explicated in previous chapters. First, like the biocentrist, I find it extremely plausible that nonsentient organisms have a welfare in the same sense that

---

141. For a discussion of one way of carving up interests in terms of their being basic, peripheral, etc. see VanDeVeer 1979.

I do, that the life of plants can go well or poorly, that they can be harmed or benefited. We attribute biological welfare to nonsentient organisms all the time when we talk about what is or isn't good for plants. These attributions don't seem to be derivative in any way, much as when we recognize that weedkiller is bad for weeds but good for us. There is a plausible account of how to make sense of the welfare of such organisms that is nonarbitrary and satisfies plausible restrictions on an account of welfare, e.g., that welfare be subject-relative. Furthermore, there are other contexts where biological welfare seems to be important, such as in medicine. Doctors make assessments about biological health and do so independently of patient's mental states: there is a conception of being a biologically healthy human being even for patients who have no desire or take no positive attitude toward being biologically healthy. Attributions of welfare to nonsentient organisms and attributions of welfare that don't seem to depend on anything cognitive or mentalistic are just part of the texture of our everyday experience. Given that I take the arguments in previous chapters to be cogent, that the most plausible account of such attributions is an etiological account of teleological welfare, and that there is no way to exclude other teleologically organized entities, I am willing to accept the consequences of those arguments.

As they say, one person's modus ponens is another's modus tollens. That accepting that there is a constituent of welfare grounded in teleology would yield the conclusion that artifacts and strange collectives have a welfare is a reason to reject the claim that this kind of welfare is genuine.[142] If we are forced to accept some counterintuitive implication, why not accept the implication that plants just don't have a welfare independently of how we typically conceive of them or what we typically attribute to them? Instead, we should to try to explicate these claims about biological welfare in some other way, to say that we must revise how we talk about nonsentient organisms or biologically healthy humans. Perhaps we can devise an understanding of "biological health" that accounts for our attributions of biological health but that is derivative or pragmatically useful but does not commit us to endorsing that biological health is a component or form of welfare. Furthermore, this approach allows us to preserve the correlate of welfarism, according to which all welfare has moral value or ought figure into the moral deliberations of the relevant

---

142. There has been some debate about whether a large range of "good for" attributions are univocal or using "good for" in the same sense. Kraut (2009) has argued that "good for" is univocal, while others have rejected that claim (Behrends 2011; Rosati 2009). Korsgaard (2014) also distinguishes senses of "good for," but she seems less committed to saying that some "good for" attributions are not welfare attributions.

agents. We need not recognize that there are two classes of beings with welfare: those that have a welfare that matters and those that have a welfare that does not.

By my lights the theoretical costs presented in favor of rejecting that there is a genuine constituent of welfare grounded in teleology are not very large. In the case of artifacts, it does seem strange to talk about most artifacts living a good or bad life, but this is simply because artifacts aren't, typically, alive. In cases where they are alive, such as synthetic organisms, it is extremely plausible to talk about their lives as going well or poorly. Furthermore, it doesn't seem strange to talk about what is good or bad for an artifact: it is good for a knife to remain sharp, it is good for a clock to keep time, etc. It is true, as discussed in chapter 5, that the explanation for why these things are good for these artifacts depends on intentions in some way, but we also saw that this doesn't undermine these things being good *for* the artifact itself. So while there are some ascriptions of welfare to artifacts that seem counterintuitive, not all such ascriptions are so.

What of the correlate of welfarism? By my lights this is not intuitive at all, and so rejecting it is not a cost. As O'Neill (2003) has argued, it seems perfectly plausible that something might promote the good of an entity, i.e., make it better off, while its promotion is not good, i.e., its promotion doesn't contribute positive value to a state of affairs. It seems perfectly reasonable to inquire about whether something's being welfare-promoting necessarily has moral value. The case for the correlate of welfarism must be made before we accept having to give it up as a genuine cost. Even if we think that there is nothing good if there is nothing good for, it just seems false that all things that are good for are good.

These considerations are not decisive. As noted, what exactly the implications of the above arguments are will be a function of balancing the reasons we have for and against the different available positions. The balance of reasons seems most clearly to show that teleocentrism is false. Beyond that, I favor accepting the etiological account of teleological welfare (or something close to it) and rejecting the correlate of welfarism and other reasons in favor of thinking that teleo-interests are not *really* a kind of welfare interest.

CONCLUSION

The primary aim of this work has been to show that biocentrism is false by developing the strongest, most plausible version of the view and then exposing it to new criticisms, criticisms that are not susceptible to the standard biocentrist responses. What are the broader implications of this project? What does it mean for those concerned with the environment or with other domains where biological welfare has seemed important? This concluding chapter aims to explore some of the most important implications of the death of the ethic of life.

Implications for the Environment

Biocentrism is a view that was developed, primarily, in the context of environmental ethics for determining, precisely, our obligations to the natural world. So, one would imagine that the implications of having to abandon biocentrism have been carefully articulated, that we already knew what was at stake in the debate. However, this isn't exactly true. While biocentrists have developed arguments that biocentrism has determinable consequences (Carter 2005; Attfield 2005; Varner 1998, chap. 4; Agar 2001, chap. 8), they haven't much articulated how environmental policy or how our general treatment of the environment should differ if, specifically, biocentrism, as opposed to sentientism, for example, is the right view of who or what has moral status.[143]

---

143. Issues of the implications about differing views of moral status have largely been about whether there can be consensus between individualists and holists (see, for example, Varner 2003, 1998; Callicott 1980) or about whether there is a form of anthropocentrism that generates obligations similar to those claimed by those with more expansive views (see, for example, the debates over Norton's [1986, 1994] "convergence hypothesis").

I think there are good reasons for this. Some of these are merely historical—Taylor, for example, takes as his primary foil the anthropocentrist, and so, while he does develop the implications of his biocentrism, there is no need for him to articulate what practically follows from accepting that all nonsentient organisms have moral status as opposed to accepting that only sentient beings have moral status.

Another reason is that the stakes are obscured by the nature of the biotic community. Humans and sentient beings all depend crucially on nonsentient organisms (and ecosystems and other biotic communities). Even if the most naive version of Anthropocentrism is true, we still have very strong obligations grounded in indirect moral status to preserve nonsentient organisms and the environments which they need to thrive. Environmental ethicists have argued that no matter one's view about moral status, whether one is a Sentientist, an Anthropocentrist, or a Holist, there are grounds for convergence over important environmental policies or practices, such as hunting (Varner 1998, chap. 5). This is why even where biocentrists have appealed to important practical implications of accepting the moral status of nonsentient organisms (Stone 1972), there are grounds for skepticism. There might be great reasons to extend rights to trees or biotic communities or act as if they have moral status that is not ultimately grounded in the actual moral status of such beings. Just as there might be very good reasons to be a pragmatic holist, there may be such reasons to be a pragmatic biocentrist. And, just as utilitarianism might be self-effacing such that the utilitarian thing to do is to convince everyone that utilitarianism is false and some other normative theory is true, anthropocentrism (or sentientism) might be self-effacing such that we should convince everyone to be a biocentrist even though the view is actually false.

The stakes of the debate over biocentrism are further obscured by the fact that most societies, certainly those that have the most environmental impact, have failed to live up to our environmental obligations, whichever among the plausible views about moral status (and normative theory) is true. It might seem that, in light of our failures, we have a duty of restitution and that this is precisely where the truth or falsity of biocentrism really matters. After all, the requirements of restitution surely differ depending on who or what has moral status. I think this is mistaken for two reasons. First, I've argued elsewhere that there are two components of our restitutive requirements: reparative requirements and remediative requirements (Basl 2010). I argued that given the nature of who is harmed, it is likely not possible to make reparations; the individuals who are wronged by our degradation, in many cases, no longer exist. Instead, I argue that meeting our obligations to make restitution, at least in the context of environmental restoration, is to be undertaken by remediating the character dispositions that yield environmental wrongdoing.[144] Second, at this point, even on an enlightened but very pro-human view of moral status, we are in the midst of an environmental crisis. If we were responsive to this crisis in appropriate ways, it would most likely involve doing most of the same actions that biocentrism or sentientism would prescribe. Perhaps how ideal environmental citizens ought to act differs according to whether biocentrism or some alternative is true, but for us very nonideal citizens, those differences seem to me to be very remote.

This could change. Given our technological prowess, we might confront issues where it is easier to delineate the consequences of biocentrism from those of sentientism or anthropocentrism. On our own planet, perhaps if we develop sufficient ability to geoengineer the climate, we will become less dependent on large swaths of nonsentient life. We are a long way from being able to replace the essential resources of Earth's plant life, but even if this were a technological possibility, there are plenty of moral reasons to avoid this path, some grounded in challenges to geoengineering and technological hubris (Gardiner 2011), and plenty more grounded in the fact that we as humans value the natural world and wish to preserve it for reasons other than those based in biocentrism.

Ironically, the stakes of the debate over biocentrism are perhaps clearest in the context of traveling to new planets. Many technologists and scientists, such as Elon Musk and Stephen Hawking, have urged that we must discover habitable planets and develop the means to travel to them because we are unlikely to survive as a species on this planet (largely because of our resource use and poor environmental management). The possibility of contaminating other planets and impacting native organisms is a live issue for NASA (Frank 2015). We can easily imagine arriving at a habitable planet containing only nonsentient life, perhaps only microbes, and having to make decisions that might jeopardize these organisms. Here we can see pretty easily what rejecting biocentrism means for us: jeopardizing these organisms itself gives rise to no obligation whatsoever to avoid colonization or the destruction of these organisms. Of course, there may be other reasons to object to colonization or the death of these organisms. It isn't that we are all things considered justified in seeking to colonize other planets or even a particular planet hosting only nonsentient life, only that these beings don't have a welfare that imposes an obligation upon us.

---

144. For criticisms of this approach, see Lintott 2011; Almassi 2017.

Implications for Biomedical Ethics and Medicine

This work has primarily focused on issues internal to environmental ethics, but appeals to biological well-being or biological health are also common in medical contexts. It is worth mentioning a few potential implications for the domain of medicine and some issues in biomedical ethics.

If biological well-being is not normative, then health assessments are, by themselves, irrelevant to medical decision-making. However, it seems clear that physicians and other medical professionals make assessments of biological health and often seem to do so independently of how this biological health figures into other, for example, mentalistic, dimensions of welfare. It is less clear whether (a) medical professionals generally take biological health to be normative and (b) how their decision making would differ if they recognized that biological health is not normative. If medical professionals don't see biological health as normative, or if biomedical health is so tightly aligned with the normative constituents of welfare, then this work has no implications in the field of medical practice and decision-making.

The question of whether medical decision-makers explicitly or implicitly believe or act as if biological health is normative is largely an empirical, sociological question. Having taught biomedical ethics many times to students training to be health professionals and having them discuss issues of biological health, it would surprise me if no medical practitioners took biological health to be normative, but this is admittedly pretty weak evidence.

What about the alignment of biological welfare and the normative constituents of welfare? Is the medical context, like the environmental context, such that there will generally be convergence between what is ultimately normative and what promotes biological welfare? For the most part, I think the answer is yes. When medical practitioners do things that improve our biological health, they almost always do things that make our life go better along some other dimension of welfare that is plausibly normative; it makes life more enjoyable, satisfies a preference, enables greater autonomy, etc.

There are cases and theories of welfare on which biological health and other dimensions of welfare come apart. Sometimes these are the kinds of cases used to motivate both the idea that there really is a biological constituent of welfare and the idea that it is normative. For example, Varner's cases of Maude and the mariners are like this. We are supposed to judge that Maude and the mariners' lives would go better under certain conditions (quitting smoking or getting vitamin C) that cannot be explained by appeal to other constituents of welfare. So there must be a biological constituent of welfare.

Furthermore, since Varner seems to accept the correlate of welfarism, these cases help to show that biological welfare is normative.

I won't rehash what I think of these as arguments for the normativity of biological welfare except to say that there are many plausible theories of welfare on which Maude and the sailors have overlapping biological and other (plausibly normative) interests that can explain not only why it would be good for Maude to stop smoking and the sailors to have an orange, but also why we have a reason to promote those ends without appealing to the normativity of biological welfare. However, even if the above arguments are sound, there are still implications for medical practice only if medical practitioners would have any reason to believe they are in a context where, for example, their patient who is a smoker has an ideal desire to smoke, or their patient, a sailor, is incapable of forming a desire or an ideal desire to have a sufficient level of vitamin C.

There are also cases medical professionals face where biological welfare potentially conflicts with other obligations (whether they are grounded in welfare or not). For those who believe that there are some instances where physician-assisted suicide is permissible, those instances will be cases where it is permissible to undermine biological health. But those who oppose all forms of euthanasia and physician-assisted suicide do not, to my knowledge, appeal to the normativity of biological health to explain or justify their opposition. So, again, these sorts of cases don't really provide any reason to worry about the implications of the death of the ethic of life in medical contexts.

However, as in environmental contexts, there are edge cases where biological health being nonnormative does have real implications. Consider body identity integrity disorder (BIID) (Giummarra et al. 2008).[145] Those suffering from BIID do not recognize some biological part of their body as a part of them. For example, patients with BIID might see their hands as not a part of them, as an invader. Whereas patients with phantom limbs identify or feel as part of them something that is not present, those with BIID identify something that is present as not a part of them. Many of those who have BIID wish to be rid of those parts they do not identify with.

It will be unsurprising that patients with BIID are typically denied requests to remove body parts to realize identity integrity. Instead, patients who seek

---

145. For a discussion of BIID and other similar disorders as well as a discussion of the ethics of voluntary amputation from a perspective that assumes a picture on which biological health is not normative, see Bayne and Levy 2005; see also Savulescu 2007.

treatment are typically treated via psychotherapy, medication, or some combination of the two. However, if biological welfare is not normative, we might ask whether medical professionals should assist patients with BIID to achieve identity integrity. In fact, to the extent that we understand "disorder" as a normative rather than purely descriptive term, we might wonder whether BIID should even count as a disorder if biological health is not normative.

Even though I think that biological health cannot ultimately ground a general policy of denying amputation or removal of body parts to patients that have BIID, there are, I think, good reasons for such a policy. One reasons is that amputation may ultimately decrease the quality of life of patients in a way that outweighs whatever gain they see in virtue of having identity integrity. Society is not always designed in ways that accommodate those with physical disabilities or those who have been dismembered, and they may face certain forms of discrimination. Another reason is that if a patient's BIID can be addressed via less invasive means, then perhaps there is some precautionary principle that justifies medical practitioners in their prescribing psychotherapy or pharmaceutical intervention before considering removal of body parts. A final reason is that it might impose on third parties a requirement that they do something they take to be unethical.

Still, the arguments of this book have made me question how general such a general policy should be. If patients suffer very greatly from BIID, if other interventions fail despite patients' best effort, and if it seems that living in a nonideal society would generally be better for patients' (nonbiological) welfare, there seems to me no very strong reason, at least not one grounded in facts about the patients themselves, to deny the person the means to identity integrity. It seems to me that if one is to ground any sort of strong prohibition against this, it must appeal to the normativity of biological welfare. Notice that it won't work to appeal to the requirement that medical professionals do no harm. That requirement is, at least presumably, grounded in an assumption that harms are normative, but my point is that biological welfare is not normative; it's not the type of harm that medical professionals have an obligation not to cause.

Since biological welfare is not normative, medical professionals and bioethicists should tread lightly in the case of BIID with respect to what treatments can be justified. Furthermore, it is worth carefully evaluating whether it should be counted as a disease. Fully developing this thought would require a defense of some account of disease or disorder. I do not intend to develop such an account or defend one that currently exists. On some accounts of disease or disorder, people have a disease or disorder just in case some part of them isn't properly functioning. But there are also normative notions of disease, on which a thing's being a disease or being recognized as such provides some reason to address this malfunction (Kingma 2014). On nonnormative accounts of disease or disorder, BIID is properly classified as a disorder (on some reasonable assumptions about the function of various traits), but on the latter accounts its status as a disease or disorder is questionable.

Emerging Technologies

Emerging technologies can make the stuff of philosophical thought experiments the stuff of our daily lives, making real to those outside philosophy what once seemed to be of concern only to the ivory tower. Consider, for example, the increased public interest in Trolley problems now that self-driving cars are within reach. At the same time, emerging technologies can give us a new context, new cases, new ideas with which to revise our philosophical frameworks. The development of AI and machine learning, for example, provides a context to think about the plausibility of the distinction between living and nonliving and its moral salience.

Emerging technologies interact with philosophical ethics in another way: they quite often force us to revisit ethical issues and give us an opportunity to draw on well-developed philosophical resources or to develop new ones to address the ethical challenges they raise. The arguments developed here, I think, help settle some important questions that are already being and will continue to be raised about artificially intelligent and autonomous systems. There are already those who are concerned with whether robots might have rights (Gunkel 2014), and some work has been done on the moral status of conscious artificial intelligences (Floridi and Sanders 2004; Coeckelbergh 2010; Basl 2013, 2014; Gunkel 2014; Schwitzgebel and Garza 2015). As robots become more autonomous, as they begin to appear as if they are making their own decisions, the question might arise as to whether these autonomous systems are morally considerable or have some other form of moral status.

The death of the ethic of life also entails that until autonomous systems are, at least, conscious or sentient, they will not be morally considerable. Unconscious systems or mere machines (Basl 2014) have a welfare, but it is of no normative importance to agents except indirectly. There will, I think, be serious epistemic hurdles to telling whether a machine is a mere machine or a fully conscious system, at least once the autonomous systems are advanced enough (Basl 2013). This might even serve as an argument not to develop such systems (Schwitzgebel and Garza 2015), but whatever restrictions there are on the development of such systems, they will not be based on the moral considerability of mere machines.

This isn't to say that advanced, but nonconscious, artificial intelligences won't have some kind of moral status. They will certainly have indirect status and might have intrinsic subjective value. An artificial intelligence that passed the Turing test might not be conscious, but it would be a wonderous thing, worthy of preservation and marvel. It's just that being impressive is an unimpressive or uninteresting way of coming to matter from the moral point of view.

Implications for Philosophy

In the practical domain, the death of the ethic of life seems to have few consequences except in certain edge cases. This raises the question of whether the question of its truth and its underpinnings even much matters. Couldn't we have known in advance that the implications of accepting biocentrism would make little difference and just set aside the question of whether biocentrism is true? In other words, is there any real payoff from this work beyond the guidance it might provide in the edge cases discussed above?

It will perhaps surprise no one that I think the answer is yes. The practical implications of philosophical work are just one dimension of the way such work pays dividends. Other dimensions include the way philosophical work clarifies disagreements, furthers changes or closes off debate, furnishes explanations (or furnishes deeper or better explanations), provides new ways for deciding between views, illuminates what questions must be settled to decide between positions, etc. In closing, I'd like to highlight some of what I hope is the philosophical payoff of this work.

First and foremost, the goal of this work has been to end a stalemate between biocentrists and sentientists (or anthropocentrists) in favor of the latter. As has been noted, parties to the debate between these views have largely seen their disagreement in terms of a difference over which beings can be said to have a welfare. My aim has been to take up and develop the theory of welfare that best serves the biocentrists and show that their position is untenable even once we grant that there is a perfectly legitimate way to make sense of, to ground, the welfare of nonsentient organisms. Furthermore, since biocentrism as defended within the welfare approach is, in my view and the view of many biocentrists, the best, the most plausible, form of biocentrism, the arguments that end the stalemate between biocentrism and sentientism within that approach also spell the death of biocentrism more generally.

Of course, I'm not so naive as to think that my arguments will convince all the targets of my arguments, some of whom have been developing their versions of biocentrism, sentientism, or a favored theory of well-being for longer than I have been doing philosophy. Those biocentrists who disagree with the conclusion of this book have several avenues of response. They may, for example,

1. Deny that the best form of biocentrism is one on which nonsentient organisms are morally considerable.
2. Deny that the welfare of nonsentient organisms is best explained in terms of teleology.
3. Deny that teleology should be grounded in selection etiologies.
4. Deny that there is no morally relevant difference between artifacts and organisms or between collectives and individuals.
5. Accept teleocentrism.

Each of these strategies (and many more) is open to the biocentrist, but each comes with significant burdens. For example, even if one were to deny that biocentrism is grounded in moral considerability, or deny that the welfare of nonsentients should be understood teleologically, or deny that teleology is best grounded in selection, one must develop completely new strategies and arguments for defending the view that conform to various constraints discussed so far; the developed view must yield the conclusion that only individual organisms (or at least only certain kinds of collectives) end up having a welfare or moral status, for example. If it turns out that the best account of the moral status of nonsentient organisms also applies, again, to artifacts, then these alternatives are of no help to the biocentrist. Indeed, this is one of the main challenges for holists, that their accounts of the intrinsic value of ecosystems are not sufficiently discriminatory.

Alternatively, some may wish to accept and defend teleocentrism; maybe this work will be cited by advocates of machine or artifact rights as just another instance of failing to recognize the ways in which moral lines should always have been drawn. Perhaps budding teleocentrists can make good on a defense of the correlate of welfarism, i.e., explain why all welfare has moral value. At the very least, proponents of teleocentrism would need to develop a nuanced view about how to trade interests against one another when those interests are of very different kinds or adapt work that has already been done on how to manage decisions where values in conflict are incommensurable. Here they might borrow from the work of the biocentric consequentialists.

Whatever strategy those who disagree adopt, they will, I think, be forced to draw on new philosophical and extraphilosophical resources, develop new inter- and intradisciplinary connections, and clarify extant views. Even if my arguments are shown to be unsound and the ethic of life thus survives unscathed, playing even a small role in advancing these new philosophical explorations would be, for me, sufficient return on investment.

WORKS CITED

Adams, Douglas. 1995. *The Hitchhiker's Guide to the Galaxy*. New York: Del Rey.

Agar, Nicholas. 1997. "Biocentrism and the Concept of Life." *Ethics* 108 (1): 147–68. https://doi.org/10.2307/2382092.

———. 2001. *Life's Intrinsic Value: Science, Ethics, and Nature*. New York: Columbia University Press.

Almassi, Ben. 2017. "Ecological Restorations as Practices of Moral Repair." *Ethics & the Environment* 22 (1): 19–40.

Amon, L. E. S., A. Beverly, and J. N. Dodd. 1984. "The Beverly Clock." *European Journal of Physics* 5 (4): 195. https://doi.org/10.1088/0143-0807/5/4/002.

Amundson, Roy, and George Lauder. 1994. "Function without Purpose." *Biology and Philosophy* 9 (4): 443–69.

Aristotle. 1999. *Nicomachean Ethics*. 2nd ed. Indianapolis: Hackett.

Arneson, Richard J. 2002. "The End of Welfare As We Know It? Scanlon versus Welfarist Consequentialism." *Social Theory and Practice* 28 (2): 315–36.

Attfield, Robin. 1981. "The Good of Trees." *Journal of Value Inquiry* 15 (1): 35–54.

———. 1995. *Value, Obligation, and Meta-ethics*. Amsterdam: Rodopi.

———. 2005. "Biocentric Consequentialism and Value-Pluralism: A Response to Alan Carter." *Utilitas* 17 (1): 85–92.

———. 2012a. "Biocentrism and Artificial Life." *Environmental Values* 21 (1): 83–94.

———. 2012b. "Synthetic Biology, Deontology and Synthetic Bioethics." *Ethics, Policy and Environment* 15 (1): 29–32.

———. 2014. *Environmental Ethics: An Overview for the Twenty-First Century*. 2nd ed. Cambridge: Polity.

Axelrod, R., and W. D. Hamilton. 1981. "The Evolution of Cooperation." *Science* 211 (4489): 1390–96.

Barker, Matthew J. 2010. "Specious Intrinsicalism." *Philosophy of Science* 77 (1): 73–91.

Barras, Colin. 2016. "The Evolution of the Nose: Why Is the Human Hooter so Big?" *New Scientist*, March 24. https://www.newscientist.com/article/2082274-the-evolution-of-the-nose-why-is-the-human-hooter-so-big/.

Basl, John. 2010. "Restitutive Restoration: New Motivations for Ecological Restoration." *Environmental Ethics* 32 (2): 135–47.

———. 2011. "The Levels of Selection and the Functional Organization of Biotic Communities." University of Wisconsin–Madison.

———. 2013. "What to Do about Artificial Consciousness." In *Ethics and Emerging Technologies*, edited by Ronald L. Sandler, 380–92. New York: Palgrave Macmillan.

———. 2014. "Machines as Moral Patients We Shouldn't Care about (Yet): The Interests and Welfare of Current Machines." *Philosophy and Technology* 27 (1): 79–96.

———. 2017. "A Trilemma for Teleological Individualism." *Synthese* 194 (4): 1057–74.

———. n.d. "Extensionism and the Levels of Selection."

Basl, John, and Ronald Sandler. 2013a. "The Good of Non-sentient Entities: Organisms, Artifacts, and Synthetic Biology." *Studies in History and Philosophy of Science Part C: Studies in History and Philosophy of Biological and Biomedical Sciences* 44 (4): 697–705.

———. 2013b. "Three Puzzles Regarding the Moral Status of Synthetic Organisms." In *Synthetic Biology and Morality: Artificial Life and the Bounds of Nature*, edited by Gregory E. Kaebnick and Thomas H. Murray, 89–106. Cambridge, MA: MIT Press.

Baxter, William. 1974. *People or Penguins: The Case for Optimal Pollution*. New York: Columbia University Press.

Bayne, Tim, and Neil Levy. 2005. "Amputees by Choice: Body Integrity Identity Disorder and the Ethics of Amputation." *Journal of Applied Philosophy* 22 (1): 75–86.

Behrends, Jeff. 2011. "A New Argument for the Multiplicity of the Good-for Relation." *Journal of Value Inquiry* 45 (2): 121–33.

Bengson, John. 2010. "The Intellectual Given." *Mind* 124 (495): 707–60.

———. 2013. "Experimental Attacks on Intuitions and Answers." *Philosophy and Phenomenological Research* 86 (3): 495–532.

Bickhard, Mark H. 2000. "Autonomy, Function, and Representation." *Communication and Cognition-Artificial Intelligence* 17 (3–4): 111–31.

Bigelow, J., and R. Pargetter. 1987. "Functions." *Journal of Philosophy* 84 (4): 181–96.

Boorse, Christopher. 1976. "Wright on Functions." *Philosophical Review* 85 (1): 70–86.

———. 1977. "Health as a Theoretical Concept." *Philosophy of Science* 44 (4): 542–73.

———. 2002. "A Rebuttal on Functions." In *Functions: New Essays in the Philosophy of Psychology and Biology*, edited by André Ariew, Robert Cummins, and Mark Perlman, 63–112. New York: Oxford University Press.

Bostrom, Nick. 2016. *Superintelligence: Paths, Dangers, Strategies*. Reprint ed. New York: Oxford University Press.

Bradshaw, John. 2012. *Dog Sense: How the New Science of Dog Behavior Can Make You a Better Friend to Your Pet*. New York: Basic Books.

Brandon, Robert N. 1995. *Adaptation and Environment*. Princeton, NJ: Princeton University Press.

———. 2013. "A General Case for Functional Pluralism." In *Functions: Selection and Mechanisms*, edited by Philippe Huneman, 97–104. Dordrecht: Springer. https://doi.org/10.1007/978-94-007-5304-4_6.

Brandt, Richard B. 1979. *A Theory of the Good and Right*. New York: Oxford University Press.

Brown, Campbell. 2011. "Consequentialize This." *Ethics* 121 (4): 749–71.

Buchanan, Allen. 2009. "Moral Status and Human Enhancement." *Philosophy & Public Affairs* 37 (4): 346–81.

———. 2011. *Beyond Humanity? The Ethics of Biomedical Enhancement*. New York: Oxford University Press.

Cahen, Harley. 2002. "Against the Moral Considerability of Ecosystems." In *Environmental Ethics: An Anthology*, edited by Andrew Light and Holmes Rolston III, 114–28. Oxford: Blackwell.

Callicott, J. Baird. 1980. "Animal Liberation: A Triangular Affair." *Environmental Ethics* 2 (4): 311–38.

———. 1990. "The Case against Moral Pluralism." *Environmental Ethics* 12 (2): 99–124.

———. 2010. "The Conceptual Foundations of the Land Ethic." In *Technology and Values: Essential Readings*, edited by Craig Hanks, 438–53. Wiley-Blackwell.

Cappelen, Herman. 2013. *Philosophy without Intuitions*. New York: Oxford University Press.

Carruthers, Peter. 2010. "Introspection: Divided and Partly Eliminated." *Philosophy and Phenomenological Research* 80 (1): 76–111.

———. 2011. *The Opacity of Mind: An Integrative Theory of Self-Knowledge*. New York: Oxford University Press.

Carter, Alan. 2005. "Inegalitarian Biocentric Consequentialism, the Minimax Implication and Multidimensional Value Theory: A Brief Proposal for a New Direction in Environmental Ethics." *Utilitas* 17 (1): 62–84.

Chalmers, David J. 2014. "Intuitions in Philosophy: A Minimal Defense." *Philosophical Studies* 171 (3): 535–44.

Coeckelbergh, Mark. 2010. "Robot Rights? Towards a Social-Relational Justification of Moral Consideration." *Ethics and Information Technology* 12 (3): 209–21.

Colyvan, Mark, Damian Cox, and Katie Steele. 2010. "Modelling the Moral Dimension of Decisions." *Noûs* 44 (3): 503–29.

Crespi, Bernard J., and Douglas Yanega. 1995. "The Definition of Eusociality." *Behavioral Ecology* 6 (1): 109–15.

Cronon, William. 1996. "The Trouble with Wilderness: Or, Getting back to the Wrong Nature." *Environmental History* 1 (1): 7–28.

Cross, Adam T., David J. Merritt, Shane R. Turner, and Kingsley W. Dixon. 2013. "Seed Germination of the Carnivorous Plant Byblis Gigantea (Byblidaceae) Is

Cued by Warm Stratification and Karrikinolide." *Botanical Journal of the Linnean Society* 173 (1): 143–52. https://doi.org/10.1111/boj.12075.

Cummins, Robert. 1975. "Functional Analysis." *Journal of Philosophy* 72: 741–64.

Damasio, Antonio. 2005. *Descartes' Error: Emotion, Reason, and the Human Brain.* Reprint ed. London: Penguin.

Daniels, Norman. 1979. "Wide Reflective Equilibrium and Theory Acceptance in Ethics." *Journal of Philosophy* 76 (5): 256–82.

Darwin, Charles. 1964. *On the Origin of Species.* Fascimile of the first edition. Cambridge, MA: Harvard University Press.

Davidson, Donald. 1987. "Knowing One's Own Mind." *Proceedings and Addresses of the American Philosophical Association* 60.

Dawkins, Richard. 1989. *The Selfish Gene.* New York: Oxford University Press.

———. 1999. *The Extended Phenotype: The Long Reach of the Gene.* New York: Oxford University Press.

Delancey, C. 2004. "Teleofunctions and Oncomice: The Case for Revising Varner's Value Theory." *Environmental Ethics* 26 (2): 171–88.

de Marneffe, Peter. 2003. "An Objection to Attitudinal Hedonism." *Philosophical Studies* 115 (2): 197–200.

Dennett, Daniel C. 1995. *Darwin's Dangerous Idea.* New York: Simon and Schuster.

Domingos, Pedro. 2015. *The Master Algorithm: How the Quest for the Ultimate Learning Machine Will Remake Our World.* New York: Basic Books.

Dreier, James. 2002. "Meta-ethics and Normative Commitment." *Noûs* 36 (s1): 241–63.

Dretske, Fred. 1995. *Naturalizing the Mind.* Cambridge, MA: MIT Press.

Dussault, Antoine C., and Frédéric Bouchard. 2016. "A Persistence Enhancing Propensity Account of Ecological Function to Explain Ecosystem Evolution." *Synthese*, May, 1–31. https://doi.org/10.1007/s11229-016-1065-5.

Elliot, Robert. 1982. "Faking Nature." *Inquiry* 25 (1): 81–93.

Engels, Mylan, Jr. 2012. "The Commonsense Case against Animal Experimentation." In *The Ethics of Animal Research: Exploring the Controversy*, edited by Jeremy R. Garrett, 215–36. Cambridge, MA: MIT Press.

Evans, Harry C., Simon L. Elliot, and David P. Hughes. 2011. "Hidden Diversity behind the Zombie-Ant Fungus Ophiocordyceps Unilateralis: Four New Species Described from Carpenter Ants in Minas Gerais, Brazil." *PLOS ONE* 6 (3): e17024. https://doi.org/10.1371/journal.pone.0017024.

Feinberg, Joel. 1980. *Rights, Justice, and the Bounds of Liberty: Essays in Social Philosophy.* Princeton, NJ: Princeton University Press.

Feldman, Fred. 2002. "The Good Life: A Defense of Attitudinal Hedonism." *Philosophy and Phenomenological Research* 65 (3): 604–28.

———. 2004. *Pleasure and the Good Life: Concerning the Nature, Varieties and Plausibility of Hedonism.* New York: Oxford University Press.

Floridi, Luciano. 2002. "On the Intrinsic Value of Information Objects and the Infosphere." *Ethics and Information Technology* 4 (4): 287–304.

Floridi, Luciano, and Jeff W. Sanders. 2004. "On the Morality of Artificial Agents." *Minds and Machines* 14 (3): 349–79.

Foot, Philippa. 2003. *Natural Goodness.* New York: Oxford University Press.

Forber, Patrick, and Rory Smead. 2014. "The Evolution of Fairness through Spite." *Proceedings of the Royal Society of London B: Biological Sciences* 281 (1780): 20132439. DOI: 10.1098/rspb.2013.2439.

Frank, Adam. 2015. "Is It Moral to Explore, and Colonize, Mars?" NPR.org, November 3, 2015. http://www.npr.org/sections/13.7/2015/11/03/454180154/is-it-moral-to-explore-and-colonize-mars.

Gardiner, Stephen M. 2011. *A Perfect Moral Storm: The Ethical Tragedy of Climate Change.* New York: Oxford University Press.

Gibson, D. G., J. I. Glass, C. Lartigue, V. N. Noskov, R. Y. Chuang, M. A. Algire, G. A. Benders, M. G. Montague, L. Ma, and M. M. Moodie. 2010. "Creation of a Bacterial Cell Controlled by a Chemically Synthesized Genome." *Science* 20 (May): 1190719. http://science.sciencemag.org/content/early/2010/05/20/science.1190719.

Gill, Steven R., Mihai Pop, Robert T. DeBoy, Paul B. Eckburg, Peter J. Turnbaugh, Buck S. Samuel, Jeffrey I. Gordon, David A. Relman, Claire M. Fraser-Liggett, and Karen E. Nelson. 2006. "Metagenomic Analysis of the Human Distal Gut Microbiome." *Science* 312 (5778): 1355–59. https://doi.org/10.1126/science.1124234.

Giummarra, Melita J., Stephen J. Gibson, Nellie Georgiou-Karistianis, and John L. Bradshaw. 2008. "Mechanisms Underlying Embodiment, Disembodiment and Loss of Embodiment." *Neuroscience & Biobehavioral Reviews* 32 (1): 143–60.

Glymour, Bruce. 2008. "Correlated Interaction and Group Selection." *British Journal for the Philosophy of Science* 59 (4): 835–55. https://doi.org/10.1093/bjps/axn033.

———. 2017. "Cross-Unit Causation and the Identity of Groups." *Philosophy of Science* 84 (4): 717–36.

Godfrey-Smith, Peter. 2000. "The Replicator in Retrospect." *Biology and Philosophy* 15 (3): 403–23. https://doi.org/10.1023/A:1006704301415.

———. 2009. *Darwinian Populations and Natural Selection.* New York: Oxford University Press.

Goode, R., and P. E. Griffiths. 1995. "The Misuse of Sober's Selection For / Selection of Distinction." *Biology and Philosophy* 10 (1): 99–108.

Goodnight, C. J., and L. Stevens. 1997. "Experimental Studies of Group Selection: What Do They Tell Us about Group Selection in Nature?" *American Naturalist* 150 (S1): 59–79.

Goodpaster, Kenneth. 1978. "On Being Morally Considerable." *Journal of Philosophy* 75: 308–25.

Gould, Stephen J. 2002. *The Structure of Evolutionary Theory.* Cambridge, MA: Belknap Press of Harvard University Press.

Gould, Stephen J., and Richard Lewontin. 1979. "The Spandrels of San Marcos and the Panglossian Paradigm." *Proceedings of the Royal Society* B 205 (1161): 581–98.

Griffin, James. 1988. *Well-Being: Its Meaning, Measurement, and Moral Importance.* New York: Oxford University Press.

Griffiths, Paul. 1993. "Functional Analysis and Proper Functions." *British Journal for the Philosophy of Science* 44 (3): 409–22.

Gunkel, David J. 2012. *The Machine Question: Critical Perspectives on AI, Robots, and Ethics.* Cambridge, MA: MIT Press.

———. 2014. "A Vindication of the Rights of Machines." *Philosophy & Technology* 27 (1): 113–32.

Hamilton, W. D. 1964a. "The Genetical Evolution of Social Behaviour. I." *Journal of Theoretical Biology* 7 (1): 1–16.

———. 1964b. "The Genetical Evolution of Social Behaviour. II." *Journal of Theoretical Biology* 7 (1): 17–52.

Hanczyc, Martin M. 2011. "Structure and the Synthesis of Life." *Architectural Design* 81 (2): 26–33.

Heathwood, Chris. 2017. "Faring Well and Getting What You Want." In *The Ethical Life: Fundamental Readings in Ethics and Moral Problems*, edited by Russ Shafer-Landau, 4th ed., 31–42. Oxford: Oxford University Press.

———. 2018. "Which Desires Are Relevant to Well-Being." *Noûs.* Online first. https://doi.org/10.1111/nous.12257.

Holm, Sune. 2012. "Biological Interests, Normative Functions, and Synthetic Biology." *Philosophy and Technology* 25 (4): 525–41.

———. 2017. "Teleology and Biocentrism." *Synthese* 194 (4): 1075–87.

Howson, Colin, and Peter Urbach. 2006. *Scientific Reasoning: The Bayesian Approach.* Chicago: Open Court.

Hull, D. L. 1980. "Individuality and Selection." *Annual Review of Ecology and Systematics* 11: 311–32.

Hursthouse, Rosalind. 1999. *On Virtue Ethics.* New York: Oxford University Press.

Jablonski, David. 1987. "Heritability at the Species Level: Analysis of Geographic Ranges of Cretaceous Mollusks." *Science* 238 (4825): 360–63.

———. 2008. "Species Selection: Theory and Data." *Annual Review of Ecology, Evolution, and Systematics* 39: 501–24.

James, Daylon, Scott A. Noggle, Tomasz Swigut, and Ali H. Brivanlou. 2006. "Contribution of Human Embryonic Stem Cells to Mouse Blastocysts." *Developmental Biology* 295 (1): 90–102.

Jamieson, Dale. 2008. *Ethics and the Environment: An Introduction.* New York: Cambridge University Press.

Joyce, R. 2002. "The Moral Value of Moss—Nicholas Agar, Life's Intrinsic Value." *Biology and Philosophy* 17 (3): 435–44.

Justus, James. 2008. "Ecological and Lyapunov Stability." *Philosophy of Science* 75 (4): 421–36.

Kaebnick, Gregory E., and Thomas H. Murray, eds. 2013. *Synthetic Biology and Morality: Artificial Life and the Bounds of Nature.* Cambridge, MA: MIT Press.

Kagan, Shelly. 1997. *Normative Ethics*. Boulder, CO: Westview Press.

Kamm, F. M. 2015. *The Trolley Problem Mysteries*. Edited by Eric Rakowski. New York: Oxford University Press.

Kant, Immanuel. 1963. "Duties to Animals and Spirits." *Lectures on Ethics*, 239–41.

Katz, Eric. 1992. "The Big Lie: Human Restoration of Nature." *Research in Philosophy and Technology* 12: 93–107.

Kerr, Benjamin, and Peter Godfrey-Smith. 2002. "Individualist and Multi-level Perspectives on Selection in Structured Populations." *Biology and Philosophy* 17 (4): 477–517.

Kingma, Elselijn. 2014. "Naturalism about Health and Disease: Adding Nuance for Progress." *Journal of Medicine and Philosophy* 39 (6): 590–608.

Kitcher, Philip, and Kim Sterelny. 1988. "The Return of the Gene." *Journal of Philosophy* 85: 553–73.

Korsgaard, Christine M. 2014. "On Having a Good." *Philosophy* 89 (3): 405–29. https://doi.org/10.1017/S0031819114000102.

Kraut, Richard. 2009. *What Is Good and Why: The Ethics of Well-Being*. Cambridge, MA: Harvard University Press.

Kripke, Saul. 1982. *Naming and Necessity*. Cambridge, MA: Harvard University Press.

Krützen, Michael, Janet Mann, Michael R. Heithaus, Richard C. Connor, Lars Bejder, and William B. Sherwin. 2005. "Cultural Transmission of Tool Use in Bottlenose Dolphins." *Proceedings of the National Academy of Sciences of the United States of America* 102 (25): 8939–43. https://doi.org/10.1073/pnas.0500232102.

Leopold, Aldo. 1966. *A Sand County Almanac*. New York: Ballantine Books.

———. 1989. *A Sand County Almanac and Sketches Here and There*. New York: Oxford University Press.

Lewens, Tim. 2004. *Organisms and Artifacts*. Cambridge, MA: MIT Press.

Lewontin, Richard. 1970. "The Units of Selection." *Annual Review of Ecology and Systematics* 1: 1–18.

Lin, Eden. 2017. "Against Welfare Subjectivism." *Noûs* 51 (2): 354–77. https://doi.org/10.1111/nous.12131.

Lintott, Sheila. 2011. "Preservation, Passivity, and Pessimism." *Ethics & the Environment* 16 (2): 95–114.

Lloyd, Elisabeth A. 1994. *The Structure and Confirmation of Evolutionary Theory*. Princeton, NJ: Princeton University Press.

———. 2007. "Units and Levels of Selection." In *The Cambridge Companion to the Philosophy of Biology*, edited by David Hull and Michael Ruse, 44–65. New York: Cambridge University Press.

———. 2015. "Adaptationism and the Logic of Research Questions: How to Think Clearly about Evolutionary Causes." *Biological Theory* 10 (4): 343–62.

Lloyd, Elisabeth A., and Stephen J. Gould. 1993. "Species Selection on Variability." *Proceedings of the National Academy of Sciences* 90 (2): 595–99.

Lovelock, J. E. 1988. *The Ages of Gaia: A Biography of Our Living Earth*. New York: Norton.

Ma, Hong, Nuria Marti-Gutierrez, Sang-Wook Park, Jun Wu, Yeonmi Lee, Keiichiro Suzuki, Amy Koski, Dongmei Ji, Tomonari Hayama, and Riffat Ahmed. 2017. "Correction of a Pathogenic Gene Mutation in Human Embryos." *Nature* 548 (7668): 413–19.

Maguire, Barry. 2015. "Grounding the Autonomy of Ethics." *Oxford Studies in Metaethics* 10: 188–215.

Mayo, Deborah G. 1996. *Error and the Growth of Experimental Knowledge*. Chicago: University of Chicago Press.

McKibben, Bill. 1999. *The End of Nature*. 10th Anniversary Ed. New York: Anchor Books.

McLoone, Brian. 2015. "Some Criticism of the Contextual Approach, and a Few Proposals." *Biological Theory* 10 (2): 116–24.

McShane, Katie. 2004. "Ecosystem Health." *Environmental Ethics* 26 (3): 227–45.

———. 2014. "Individualist Biocentrism vs. Holism Revisited." *Les ateliers de l'ethicque/The Ethics Forum* 9 (2): 130–48.

Miklósi, Adám, Enikö Kubinyi, József Topál, Márta Gácsi, Zsófia Virányi, and Vilmos Csányi. 2003. "A Simple Reason for a Big Difference: Wolves Do Not Look Back at Humans, but Dogs Do." *Current Biology* 13 (9): 763–66.

Mill, John Stuart. 2009. *Three Essays on Religion*. Orchard Park, NY: Broadview Press.

Millikan, Ruth Garrett. 1989. "In Defense of Proper Functions." *Philosophy of Science* 56 (2): 288–302.

———. 1999. "Wings, Spoons, Pills, and Quills: A Pluralist Theory of Function." *Journal of Philosophy* 96 (4): 191–206.

Millstein, Roberta L. 2009. "Populations as Individuals." *Biological Theory* 4 (3): 267–73.

Mitchell, Sandra D. 1993. "Dispositions or Etiologies? A Comment on Bigelow and Pargetter." *Journal of Philosophy* 90 (5): 249–59.

Moore, G. E. 2004. *Principia Ethica*. Mineola, NY: Dover Philosophical Classics.

Mossio, M., and L. Bich. 2014. "What Makes Biological Organisation Teleological?" *Synthese* 194 (4): 1–26.

Muir, William M. 1996. "Group Selection for Adaptation to Multiple-Hen Cages: Selection Program and Direct Responses." *Poultry Science* 75 (4): 447–58.

———. 2003. "Indirect Selection for Improvement of Animal Well-Being." In *Poultry Genetics, Breeding and Biotechnology*, edited by W. M. Muir and S. E. Aggrey, 247–56. Cambridge, MA: CABI Publishing.

Nagasawa, Miho, Shouhei Mitsui, Shiori En, Nobuyo Ohtani, Mitsuaki Ohta, Yasuo Sakuma, Tatsushi Onaka, Kazutaka Mogi, and Takefumi Kikusui. 2015. "Oxytocin-Gaze Positive Loop and the Coevolution of Human-Dog Bonds." *Science* 348 (6232): 333–36. https://doi.org/10.1126/science.1261022.

Neander, Karen. 1988. "What Does Natural Selection Explain? Correction to Sober." *Philosophy of Science* 55 (3): 422–26.

———. 1991a. "Functions as Selected Effects: The Conceptual Analyst's Defense." *Philosophy of Science* 58 (2): 168–84.

———. 1991b. "The Teleological Notion of 'Function.'" *Australasian Journal of Philosophy* 69 (4): 454–68.

———. 1995. "Pruning the Tree of Life." *British Journal for the Philosophy of Science* 46 (1): 59–80.

Norcross, A. 2004. "Puppies, Pigs, and People: Eating Meat and Marginal Cases." *Philosophical Perspectives* 18 (1): 229–45.

Norton, Bryan G. 1994. *Toward Unity among Environmentalists.* New York: Oxford University Press.

Nozick, Robert. 1974. *Anarchy, State, and Utopia.* New York: Basic Books.

Nussbaum, Martha C. 2013. *Creating Capabilities: The Human Development Approach.* Reprint ed. Cambridge, MA: Belknap Press of Harvard University Press.

Odenbaugh, Jay. 2010. "On the Very Idea of an Ecosystem." In *New Waves in Metaphysics*, edited by Allan Hazlett, 240–58. New York: Palgrave Macmillan.

———. 2017. "Nothing in Ethics Makes Sense Except in the Light of Evolution? Natural Goodness, Normativity, and Naturalism." *Synthese* 194 (4): 1031–55.

Okasha, Samir. 2002. "Darwinian Metaphysics: Species and the Question of Essentialism." *Synthese* 131 (2): 191–213.

———. 2006. *Evolution and the Levels of Selection.* New York: Oxford University Press.

O'Neill, John. 2003. "The Varieties of Intrinsic Value." In *Environmental Ethics: An Anthology*, edited by Andrew Light and Holmes Rolston III, 131–42. Oxford: Blackwell.

Parfit, Derek. 1986. *Reasons and Persons.* New York: Oxford University Press.

Powell, Russell. 2011. "On the Nature of Species and the Moral Significance of Their Extinction." In *The Oxford Handbook of Animal Ethics*, edited by Tom L. Beauchamp and R. G. Frey, 603–27. New York: Oxford University Press.

Preston, Christopher. 2008. "Synthetic Biology: Drawing a Line in Darwin's Sand." *Environmental Values* 17 (1): 23–39.

Raibley, Jason. 2010. "Well-Being and the Priority of Values." *Social Theory and Practice* 36 (4): 593–620.

Rämer, Patrick C., Obinna Chijioke, Sonja Meixlsperger, Carol S. Leung, and Christian Münz. 2011. "Mice with Human Immune System Components as in Vivo Models for Infections with Human Pathogens." *Immunology and Cell Biology* 89 (3): 408–16.

Rawls, John. 1999. *A Theory of Justice.* Cambridge, MA: Harvard University Press.

Raz, Joseph. 1986. *The Morality of Freedom.* Oxford: Clarendon Press.

Regan, Tom. 1983. *The Case for Animal Rights.* Berkeley: University of California Press.

Rice, Christopher M. 2013. "Defending the Objective List Theory of Well-Being." *Ratio* 26 (2): 196–211.

Rolston, Holmes III. 1989. *Philosophy Gone Wild*. Amherst, NY: Prometheus Books.

———. 2003. "Value in Nature and the Nature of Value." In *Environmental Ethics: An Anthology*, edited by Andrew Light and Holmes Rolston III, 143–53. Oxford: Blackwell.

Rosati, Connie S. 2009. "Relational Good and the Multiplicity Problem." *Philosophical Issues* 19 (1): 205–34.

Ross, W. D. 1988. *The Right and the Good*. Reprint ed. Indianapolis: Hackett.

Routley, Richard. 1973. "Is There a Need for a New, an Environmental Ethic." In *Proceedings of the XVth World Congress of Philosophy*, 1: 205–10. Sofia: Sofia Press.

Sample, Ian. 2017. "'It's Able to Create Knowledge Itself': Google Unveils AI That Learns on Its Own." *The Guardian*, October 18, 2017, sec. Science. http://www.theguardian.com/science/2017/oct/18/its-able-to-create-knowledge-itself-google-unveils-ai-learns-all-on-its-own.

Sandler, Ronald. 2007. *Character and Environment: A Virtue-Oriented Approach to Environmental Ethics*. New York: Columbia University Press.

Sandler, Ronald, and Luke Simons. 2012. "The Value of Artefactual Organisms." *Environmental Values* 21 (1): 43–61.

Sanz, Crickette M., Josep Call, and Christophe Boesch, eds. 2014. *Tool Use in Animals: Cognition and Ecology*. Reprint ed. New York: Cambridge University Press.

Sarkar, Sahotra. 2007. "A Note on Frequency Dependence and the Levels/Units of Selection." *Biology & Philosophy* 23 (2): 217–28. https://doi.org/10.1007/s10539-007-9092-8.

Savulescu, Julian. 2007. "Autonomy, the Good Life, and Controversial Choices." *The Blackwell Guide to Medical Ethics*, edited by Rosamond Rhodes, Leslie P. Francis, and Anita Silvers, 17–37. Oxford: Blackwell.

Scanlon, Thomas. 1998. *What We Owe to Each Other*. Cambridge, MA: Harvard University Press.

Schroeder, Mark. 2015. "Normative Ethics and Metaethics." In *Routledge Handbook of Metaethics*, edited by Tristram McPherson and David Plunkett, 674–86. New York: Routledge.

Schweitzer, Albert. 1969. *Reverence for Life*. Translated by Reginald H. Fuller. New York: Harper & Row.

Schwitzgebel, Eric, and Mara Garza. 2015. "A Defense of the Rights of Artificial Intelligences." *Midwest Studies in Philosophy* 39 (1): 98–119.

Sen, Amartya. 1993. "Capability and Well-Being." In *The Quality of Life*, edited by Amartya Sen and Martha Nussbaum, 1:30–54. New York: Oxford University Press.

Singer, Peter. 2009. *Animal Liberation: The Definitive Classic of the Animal Movement*. Reissue ed. New York: Harper Perennial Modern Classics.

Skyrms, Brian. 2014. *Evolution of the Social Contract*. 2nd ed. New York: Cambridge University Press.

Slote, Michael. 2001. *Morals from Motives.* New York: Oxford University Press.

Smith, J. Maynard. 1964. "Group Selection and Kin Selection." *Nature* 201: 1145–47.

Sobel, David. 1997. "On the Subjectivity of Welfare." *Ethics* 107 (3): 501–8.

Sober, Elliott. 1986. "Philosophical Problems for Environmentalism." In *The Preservation of Species*, edited by Bryan Norton, 173–94. Princeton, NJ: Princeton University Press.

———. 2000. *Philosophy of Biology.* Boulder, CO: Westview Press.

———. 2008. *Evidence and Evolution: The Logic Behind the Science.* New York: Cambridge University Press.

———. 2010. *Did Darwin Write the Origin Backwards: Philosophical Essays on Darwin's Theory.* Amherst, NY: Prometheus Books.

———. 2014. *The Nature of Selection: Evolutionary Theory in Philosophical Focus.* Chicago: University of Chicago Press.

———. 2015. *Ockham's Razors.* New York: Cambridge University Press.

Sober, Elliott, and David Sloan Wilson. 1998. *Unto Others: The Evolution and Psychology of Unselfish Behavior.* Cambridge, MA: Harvard University Press.

Sterba, James P. 1998. "A Biocentrist Strikes Back." *Environmental Ethics* 20 (4): 361–76.

Stone, Christopher D. 1972. "Should Trees Have Standing? Toward Legal Rights for Natural Objects." *Southern California Law Review* 45 (2): 450–501.

Streiffer, Robert. 2015. "Human/Non-Human Chimeras." In *The Stanford Encyclopedia of Philosophy*, edited by Edward N. Zalta, Winter 2015 ed. https://plato.stanford.edu/archives/win2015/entries/chimeras/.

Streiffer, Robert, and John Basl. 2011. "Ethical Issues in the Application of Biotechnology to Animals in Agriculture." In *The Oxford Handbook of Animal Ethics*, edited by Tom L. Beauchamp and R. G. Frey, 826–54. New York: Oxford University Press.

Sumner, L. W. 1995. "The Subjectivity of Welfare." *Ethics* 105 (4): 764–90.

———. 1999. *Welfare, Happiness, and Ethics.* New York: Oxford University Press.

Swanton, Christine. 2015. *The Virtue Ethics of Hume and Nietzsche.* Vol. 3. New York: John Wiley & Sons.

Swenson, William, J. Arendt, and David Sloan Wilson. 2000. "Artificial Selection of Microbial Ecosystems for 3-Chloroaniline Biodegradation." *Environmental Microbiology* 2 (5): 564–71.

Swenson, William, David Sloan Wilson, and Roberta Elias. 2000. "Artificial Ecosystem Selection." *Proceedings of the National Academy of Sciences* 97 (16): 9110–14. https://doi.org/10.1073/pnas.150237597.

Sylvan, R. 1994. *Against the Main Stream: Critical Environmental Essays.* Canberra: Australian National University.

Taylor, Paul W. 1989. *Respect for Nature.* Princeton, NJ: Princeton University Press.

Thomson, Judith Jarvis. 1971. "A Defense of Abortion." *Philosophy and Public Affairs* 1 (1): 47–66.

Throop, Bill. 1999. "Refocusing Ecocentrism." *Environmental Ethics* 21 (1): 3–21.

Toepfer, Georg. 2012. "Teleology and Its Constitutive Role for Biology as the Science of Organized Systems in Nature." *Studies in History and Philosophy of Science Part C* 43 (1): 113–19.

VanDeVeer, D. 1979. "Interspecific Justice." *Inquiry* 22 (1): 55–79.

Van Fraassen, Bas C. 1977. "The Pragmatics of Explanation." *American Philosophical Quarterly* 14 (2): 143–50.

———. 1980. *The Scientific Image.* Oxford: Clarendon Press.

Varner, Gary. 1998. *In Nature's Interest.* New York: Oxford University Press.

———. 2003. "Life's Intrinsic Value." *Environmental Ethics* 25 (4): 413–16.

———. 2012. *Personhood, Ethics, and Animal Cognition: Situating Animals in Hare's Two Level Utilitarianism.* New York: Oxford University Press.

Velasco, Joel D. 2008. "Species Concepts Should Not Conflict with Evolutionary History, but Often Do." *Studies in History and Philosophy of Science Part C* 39 (4): 407–14.

Wade, M. J. 1976. "Group Selections among Laboratory Populations of Tribolium." *Proceedings of the National Academy of Sciences* 73 (12): 4604–7.

Wakefield, Jerome C. 1992. "The Concept of Mental Disorder. On the Boundary between Biological Facts and Social Values." *American Psychologist* 47 (3): 373–88. https://doi.org/http://dx.doi.org.ezproxy.neu.edu/10.1037/0003-066X.47.3.373.

Waters, C. K. 2005. "Why Genic and Multilevel Selection Theories Are Here to Stay." *Philosophy of Science* 72 (2): 311–33.

Williams, George Christopher. 1996. *Adaptation and Natural Selection: A Critique of Some Current Evolutionary Thought.* Princeton, NJ: Princeton University Press.

Wilson, David Sloan, and William Swenson. 2003. "Community Genetics and Community Selection." *Ecology* 84 (3): 586–88.

Wilson, Edward O. 1984. *Biophilia.* Reprint ed. Cambridge, MA: Harvard University Press.

Wilson, Robert A. 2004. "Test Cases, Resolvability, and Group Selection: A Critical Examination of the Myxoma Case." *Philosophy of Science* 71 (3): 380–401.

Wouters, A. 2005. "The Function Debate in Philosophy." *Acta Biotheoretica* 53 (2): 123–51.

Wright, Larry. 1973. "Functions." *Philosophical Review* 82: 139–68.

INDEX